AF773547

Alzheimer's Disease

Alzheimer's Disease

Understanding Biomarkers, Big Data, and Therapy

AHMED A. MOUSTAFA

School of Psychology, Western Sydney University, Sydney, NSW, Australia; MARCS Institute for Brain and Behaviour, Western Sydney University, Sydney, NSW, Australia; Department of Human Anatomy and Physiology, the Faculty of Health Sciences, University of Johannesburg, Johannesburg, South Africa

ELSEVIER

ACADEMIC PRESS

An imprint of Elsevier

Academic Press is an imprint of Elsevier
125 London Wall, London EC2Y 5AS, United Kingdom
525 B Street, Suite 1650, San Diego, CA 92101, United States
50 Hampshire Street, 5th Floor, Cambridge, MA 02139, United States
The Boulevard, Langford Lane, Kidlington, Oxford OX5 1GB, United Kingdom

British Library Cataloguing-in-Publication Data
A catalogue record for this book is available from the British Library

Library of Congress Cataloging-in-Publication Data
A catalog record for this book is available from the Library of Congress

ISBN: 978-0-12-821334-6

For Information on all Academic Press publications
visit our website at https://www.elsevier.com/books-and-journals

Publisher: Nikki Levy
Acquisitions Editor: Joslyn Chaiprasert-Paguio
Editorial Project Manager: Andrea R. Dulberger
Production Project Manager: Omer Mukthar
Cover Designer: Christian J. Bilbow

Typeset by MPS Limited, Chennai, India

Dedication

To all the patients with Alzheimer's disease I have worked over the last decade. Your time and efforts have made this research possible.

Contents

List of contributors

Eid Abo Hamza
Faculty of Education, Department of Mental Health, Tanta University, Tanta, Egypt; College of Graduate Studies, Arabian Gulf University, Bahrain

Hany Alashwal
MARCS Institute for Brain and Behaviour, Western Sydney University, Sydney, NSW, Australia; College of Information Technology, United Arab Emirates University, Al-Ain, United Arab Emirates

Phoebe Bailey
School of Psychology, Western Sydney University, Sydney, NSW, Australia; MARCS Institute for Brain, Behaviour, and Development, Western Sydney University, Sydney, NSW, Australia

Lily Bilson
School of Psychology, Western Sydney University, Sydney, NSW, Australia

Mohamad El Haj
Laboratoire de Psychologie des Pays de la Loire (LPPL-EA 4638), Nantes Université, Univ Angers, Nantes, France; Unité de Gériatrie, Centre Hospitalier de Tourcoing, Tourcoing, France; Institut Universitaire de France, Paris, France

Sandra Garrido
School of Psychology, Western Sydney University, Sydney, NSW, Australia

Shimaa Adel Heikal
Biotechnology Program, School of Science and Engineering, The American University in Cairo, New Cairo, Egypt

Ahmed Helal
Faculty of Education, Department of Mental Health, Tanta University, Egypt

Wafa Jaroudi
MARCS Institute for Brain, Behaviour, and Development, Western Sydney University, Sydney, NSW, Australia; School of Social Sciences and Psychology, Western Sydney University, Sydney, NSW, Australia

Rasu Karki
School of Psychology, Western Sydney University, Sydney, NSW, Australia

Ahmed A. Moustafa
School of Psychology, Western Sydney University, Sydney, NSW, Australia; MARCS Institute for Brain and Behaviour, Western Sydney University, Sydney, NSW, Australia; Department of Human Anatomy and Physiology, The Faculty of Health Sciences, University of Johannesburg, Johannesburg, South Africa

Abdrabo Soliman
Department of Social Sciences, College of Arts and Sciences, Qatar University, Doha, Qatar

Dustin van der Haar
Academy of Computer Science and Software Engineering at the University of Johannesburg, South Africa

Samuel L. Warren
Psychological Science, School of Psychology, Western Sydney University, Sydney, NSW, Australia

Acknowledgments

I thank all students, research assistants, and collaborators who made this book possible.

I would like to personally thank Samuel L. Warren, Wafa Jaroudi, Lily Bilson, Mohamad El Haj, Rasu Karki, Eid Abo Hamza, and Sandra Garrido. It has been a pleasure collaborating with you on several projects on dementia.

PART I

Mental health in dementia

CHAPTER 1

Measures of depression in Alzheimer's disease

Ahmed A. Moustafa[1,2,3], Wafa Jaroudi[4], Ahmed Helal[5], Lily Bilson[1] and Mohamad El Haj[6,7,8]
[1]School of Psychology, Western Sydney University, Sydney, NSW, Australia
[2]MARCS Institute for Brain and Behaviour, Western Sydney University, Sydney, NSW, Australia
[3]Department of Human Anatomy and Physiology, The Faculty of Health Sciences, University of Johannesburg, Johannesburg, South Africa
[4]School of Social Sciences and Psychology, Western Sydney University, Sydney, NSW, Australia
[5]Faculty of Education, Department of Mental Health, Tanta University, Egypt
[6]Laboratoire de Psychologie des Pays de la Loire (LPPL-EA 4638), Nantes Université, Univ Angers, Nantes, France
[7]Unité de Gériatrie, Centre Hospitalier de Tourcoing, Tourcoing, France
[8]Institut Universitaire de France, Paris, France

Introduction

Depression impacts a large number of patients with dementia and Alzheimer's disease (AD), with several studies reporting between 20% and 60% of AD patients diagnosed with depression (Reichman & Coyne, 1995). Diagnosing depressive disorder in dementia is difficult due to overlap in symptoms, poor communication of symptoms, and lack of insight (Burke, Goldfarb, Bollam, & Khokher, 2019; Dias, Barbosa, Kuang, & Teixeira, 2020). Shared symptoms between depression and dementia include sleep disturbances, changes in eating behavior, decreased initiative and interest (apathy), psychomotor agitation, poor concentration, anxiety, and tearfulness. Depressive symptoms not commonly seen in dementia *without* depression are consistent sadness, marked morning mood worsening, feelings of worthlessness or excessive or inappropriate guilt, recurrent thoughts of death and suicidal ideation or behaviors.

Despite sad mood being the most common neuropsychiatric symptom in AD, there is no general consensus on how to diagnose depression in AD. The National Institute of Mental Health (NIMH) has proposed Provisional Diagnostic Criteria for Depression in Alzheimer's Disease (PDC-dAD). However, it has not yet been critically validated for research and clinical practice. Scales are often used to reduce clinician uncertainty when making a diagnosis and monitoring changes in AD. However, there have been several scales used to measure depression in dementia patients (see Table 1.1).

Alzheimer's Disease
DOI: https://doi.org/10.1016/B978-0-12-821334-6.00011-9

Table 1.1 List of most existing scales to measure depression in AD.

Scale	Symptoms considered	Comments	Relevant studies
Beck's Depression Inventory	Sadness; discouragement; feelings of failure; levels of satisfaction; feelings of guilt; punishment; disappointment; blame; suicide; crying; irritability; interest; decision making; appearance; effort; sleep; fatigue; appetite; weight; worry; libido.		Chiu et al. (2018), O'Shea et al. (2018), Zlatar, Muniz, Galasko, and Salmon (2017)
DSM-IV criteria			Starkstein et al. (2011), Vilalta-Franch et al. (2006)
Cornell Scale for Depression in Dementia	Mood (anxiety, sadness, reactivity, irritability); behavior (agitation, retardation, physical complaints, loss of interest); physical (appetite loss, weight loss, lack of energy); cyclic functions (diurnal variation of mood, difficulty falling asleep, multiple awakenings during sleep, early morning awakenings); ideation disturbance (suicide, poor self-esteem, pessimism, mood congruent delusions).	Scoring system: ratings are based on symptoms and signs occurring during the week prior to interview. No score should be given if symptoms result from physical disability or illness.	Sheehan (2012); Thorpe (2009), Barca et al. (2008), Portugal Mda et al. (2012), Muller-Thomsen et al. (2005)
Dementia Mood Assessment Scale	Motor activity; sleep (insomnia and daytime drowsiness); appetite (decreased appetite and increased appetite); psychosomatic complaints; energy; irritability; physical agitation; anxiety; depressed appearance; awareness of emotional state; emotional responsiveness; sense of enjoyment; self-esteem; guilt feelings; hopelessness/helplessness; suicidal thoughts; speech; diurnal mood variation (morning/evening and severity); diurnal cognitive variations (morning/evening and severity); paranoid		Thorpe (2009)

(*Continued*)

Table 1.1 (Continued)

Scale	Symptoms considered	Comments	Relevant studies
	symptoms; other psychotic symptoms; expressive communication skills; receptive cognitive capacity; cognitive insight.		
Geriatric Depression Scale (GDS)	Life satisfaction; dropped activities/interests; emptiness of life; boredom; good spirit; afraid of bad things happening; happiness; helpless; preferences of staying home/going out; memory problems; worthlessness; energy; situation is hopeless; life compared to other life.		
Hamilton Depression Scale	Depressed mood; guilt; suicide; insomnia (initial, middle, delayed); work and interests; retardation agitation; anxiety (psychic, somatic); somatic symptoms (gastrointestinal); genital symptoms; hypochondriasis (obsessed thinking they have medical condition undiagnosed); weight loss; insight; diurnal variation; depersonalization and derealization; paranoid symptoms; obsession symptoms.		Sheehan (2012)
Hospital Anxiety and Depression Scale	Tension; enjoyment; fright; laugh; worry; cheerful; relaxation; slowed down? Fight (butterflies)—nervousness? Interest in appearance; restless; enjoyment towards things/events.	Scale instructions: how you have been feeling in the past week?	Sheehan (2012)

(Continued)

Table 1.1 (Continued)

Scale	Symptoms considered	Comments	Relevant studies
Montgomery-Asberg Depression Scale	Apparent sadness; reported sadness; inner tension; reduced sleep; reduced appetite; concentration difficulties; lassitude; inability to feel; pessimistic thoughts; suicidal thoughts.	Sheehan (2012) comments that this scale is commonly used in intervention research, implying that it can possibly be used to compare pre and postintervention depression in dementia patients.	Portugal Mda et al. (2012), Sheehan (2012); Muller-Thomsen et al. (2005)
Nurses Observations Scale for Geriatric Patients	Appearance; activities of enjoyment; sadness; restlessness at night; interest in surroundings; tidiness; bowel control; memory; ability to go shopping; worthlessness; activity in hobbies; repetition in speech; sadness/crying/tearfulness; Hygiene-clean and tidy (including dress) appearance; runs away; remembers names; helps others to the extent of ability; orientation in usual surroundings; quarrelsome and irritable when questioned; aggressiveness (verbal/physical); urinary continence; cheerfulness; contact with friends/family; identity confusion of people; enjoys events (parties); friendly, positive; stubborn behavior.	Scale instructions: based on nurses observations—how the patient has been doing in the last 2 weeks?Nurses are instructed to consider the patient over the past 2 weeks; responding is subjective and their type of relationship and contact with the patients might, over the period of time in care, still potentially influence their responses.	Muller-Thomsen et al. (2005)

(Continued)

Table 1.1 (Continued)

Scale	Symptoms considered	Comments	Relevant studies
Patient Health Questionnaire	Loss of pleasure; feeling down, depressed, hopeless; sleep problems; fatigue; appetite change; feelings of failure or guilt; concentration troubles; slower speech, motor functioning; fidgety or restless; thoughts of suicide or self-harm.	Scale treatment and monitoring using scale: a depression diagnosis that warrants treatment or a treatment change needs at least one of the first two questions endorsed as positive in the past 2 weeks. Scale instructions: over the past 2 weeks, how often have you been bothered by any of the following problems?	Kroenke, Spitzer, and Williams (2003), Kroenke, Spitzer, Williams, and Lowe (2010)
NIMH Provisional Diagnostic Criteria for Depression in AD			Starkstein et al. (2011), Sepehry et al. (2017), Vilalta-Franch et al. (2006)
Cambridge Examination for Mental Disorder of the Elderly			Ball et al. (2004), Martinelli, Cecato, Bartholomeu, and Montiel (2014), Vilalta-Franch et al. (2006)

(*Continued*)

Table 1.1 (Continued)

Scale	Symptoms considered	Comments	Relevant studies
Neuropsychiatric Inventory			Connor, Sabbagh, and Cummings (2008), Vilalta-Franch et al. (2006)
ICD-10			Vilalta-Franch et al. (2006), Wang et al. (2018)
Dementia Mood Assessment Scale			Thorpe (2009)
Alzheimer's Mood Scale			Robert et al. (2010)
Zung self-rating depression scale	Sad; feeling best in morning; crying; trouble sleeping at night; appetite; libido; weight loss; constipation; faster heart beat; fatigue (tired); mind clearness; difficulty doing things (easy?); restlessness; decision making; usefulness and needed; life is pretty full? Suicidal thoughts; activity enjoyment.	Survey instructions: how often you felt or behaved this way during the past several days?	Gottlieb, Gur, and Gur (1988), Zung (1965), Zung, Richards, and Short (1965)

Measures of depression in AD

Many studies use several depression scales for the evaluation of depression in dementia patients (Brodaty & Luscombe, 1996; Teresi, Abrams, Holmes, Ramirez, & Eimicke, 2001); yet, very few compare and evaluate these scales in comparison to each other in terms of which better identifies depression in dementia patients or, if they measure different or similar variables of depression and dementia. As such, Knapskog, Barca, and Engedal (2013) evaluated AD patients' scores on the Cornell Scale for Depression in Dementia (CSDD) and the Montgomery-Asberg Depression Rating Scale (MADRS) to compare assessment of depression and, investigate the correlation between the scales and how this correlation was related to the variables of patients' age, gender, level of cognitive impairment, and level of caregiver burden. A total of 520 patients were involved in the Knapskog et al. (2013) study as patients recruited from memory clinics with dementia, mild cognitive impairment, or subjective cognitive impairment. The examination of patients' suitability for the study involved dementia diagnosis and consensus between physicians regarding results of the following assessments: neuropsychological examination (measured using the Mini-Mental State Examination (MMSE) and Clock Drawing Test), physical examination (evaluated using blood samples and interviews with patients or caregivers, noting observations made), anatomical examination (measured using CT or MRI brain scans, single photon emission CT scan and spinal fluid test), and International Statistical Classification of Diseases (ICD)assessment. To differentiate and evaluate participants as either having dementia, mild cognitive impairment, or subjective cognitive impairment, impairment was evaluated using measures of Activities of Daily Living (ADL) or revising family doctor referrals to memory clinics, giving physicians an idea of patients' memory problems. To more specifically evaluate patients' depression over the previous month prior to study involvement, trained nurses completed the CSDD and physicians who examined patients completed the MADRS independently of each other. The completion of the measures was done on the basis of two different interviews using two sources of interviewer information. In assessing patients' caregiver burden level, nurses completed the Relative Stress Scale on the basis of information provided by the caregivers using a rating scale of 0 ("not at all") to 4 ("a high degree") in response to statements. Moreover, Spearman's rho and principal component analysis were employed to assess the correlation between the CSDD and MADRS scale of measurement. Regarding the correlations between the CSDD and MADRS measures, correlation was poor in mild cognitive impairment and individuals with subjective cognitive impairment, and even worse in patients with dementia. Knapskog et al. (2013) suggest that the weakest correlation between the scales in dementia patients is due to their impaired memory compared to mild cognitive impairment and individuals with subjective cognitive impairment. They suggested that impaired memory may lead to less reporting of depressive symptoms in dementia patients in comparison to individuals with mild cognitive impairment and subjective cognitive impairment. In evaluation of the variables of patient characteristics influencing the correlation between the depression measures, statistical analysis overall did not reveal cognitive impairment as a factor influencing the relation between the CSDD and MADRS scales. However, patients' neuropsychological examination using the MMSE (scores between 26 and 28) was related to the correlation between the depression scales whereas the clock drawing results did not have any relation to the CSDD or MADRS scales correlation. On the other hand, the caregiver burden measure did not relate to the correlation between the CSDD and MADRS measures. Conclusions that can be drawn from this study include a poor relationship between the CSDD and MADRS scales of measurement of depression in dementia. Further, the authors suggest that depression in patients with memory problems should therefore be evaluated by using both the patient and caregiver since, as patients' memory worsens (as seen in the dementia participants sample), reporting of depressive symptoms decreases, due to memory problems.

Sheehan (2012) reviews scales in domains of cognition, function, behavior, quality of life (QOL), depression in dementia, carer burden, and overall dementia severity. Sheehan (2012) also discusses the important characteristics needed for scales to be accurate and effective measures of depression in dementia. It is very important for scales to have face validity (patient, family, and clinicians agree that the questions are relevant and important), construct validity (measures the construct it was designed to measure), concurrent validity (performs well against other well-performing measures), and reliability (interrater and test-retest). Although there are many scales that measure depression in dementia, Sheehan (2012) suggests the gold standard for quantifying depressive symptoms in those with dementia to be the CSDD. Other depression scales such as the Geriatric Depression Scale (GDS), the MADRS, and the Hospital Anxiety and Depression Scale were considered less accurate with patients suffering moderate to severe dementia. Sheehan (2012) suggested this is possibly due to those patients having difficulty comprehending the questions in these scales. The Hamilton Depression Rating Scale was also reviewed but reported to be too lengthy to be considered practical for use with dementia patients. Understanding the findings of the review therefore highlights that selecting the most accurate scale is essential for being able to clarify intervention effects in dementia patients.

Thorpe (2009) identifies two scales to diagnose depression in dementia: the Dementia Mood Assessment Scale (DMAS) and the CSDD. Thorpe (2009) also details the necessary areas to consider when collecting a careful clinical assessment of symptom history, family history, and laboratory investigations. Careful symptom history includes the following: detailed description of symptoms, their time course and progression, and their association with other confounding factors such as environmental stressors including pain, poor nutritional status, other medical conditions, and recent medication changes. Particular attention is also to be paid to depression symptoms that are *less common in dementia* such as hopelessness, expressions of guilt, feeling worthless, and thoughts of self-harm. Thorpe (2009) also suggests that prefrontal cortex-related impairments such as disinhibition, perseveration, and decreased initiative are indicative of strong frontal dementia rather than depression. Information about family history of mood disorders, previous personal history of depression, and previous responses to therapy is suggested to be important. Within the direct interview setting one should pay close attention to consistently low mood and affect that does not respond to stimulation, hopelessness, expressions of guilt, feelings of worthlessness, and thoughts of self-harm. Thorpe (2009) also suggested that laboratory investigation should also be undertaken for hematology, thyroid function, electrolytes, vitamin B12, and drug levels of medications as they all have a propensity to cause mood symptoms, such as depression, in dementia patients.

Kirkham et al. (2016) estimated that 20%–30% of people with AD will develop major depressive disorder (MDD). In order to diagnose depression in dementia, patients need to present three symptoms as opposed to five in populations without dementia. The first symptom must be either depressed mood or loss of enjoyment in activities. The remaining symptoms can be any of the following: significant change in weight or appetite, insomnia or hypersomnia, psychomotor agitation or retardation, fatigue, feelings of worthlessness or guilt, difficulties concentrating, and recurrent thoughts of death or suicide. Untreated depression in dementia is associated with increased mortality, accelerated cognitive decline, earlier nursing home placement, and decreased QOL. The objective of the Kirkham et al. (2016) study was to identify the accuracy of depression rating scales as screening tools for detecting depression in dementia and to compare their diagnostic accuracy. Its secondary objective was to examine factors that may impact the accuracy of depression rating scales. These include the reference standard used for depression in dementia (the Diagnostic and Statistical Manual), the baseline prevalence of depression in dementia, age, gender, type of dementia, study setting, and study country. Participants involved in the study were dementia patients diagnosed with AD, vascular dementia, mixed vascular and AD dementia, and dementia with Lewy bodies. The index tests include several generic depression scales as well as scales specific to depression in dementia.

Portugal Mda et al. (2012) assessed the validity of the MADRS and the CSDD in 95 patients with dementia, aged 65 years and older. They also attempted to find the cutoffs with respective sensitivity and specificity indices, which are most suitable for recognizing depression in dementia using both scales. While sensitivity is the probability of having a positive test if one has the disorder, specificity is the probability of having a negative test and not having the disorder. The GDS and the Nurses Observation Scale for Geriatric Patients (NOSGER) have not reached a consistent cutoff to recognize depression in the elderly, especially for those with dementia, making them unreliable. The MADRS and the CSDD have been considered useful as they reach a satisfactory agreement with gold standard instruments to diagnose depression in patients with dementia, regardless of severity (Muller-Thomsen, Arlt, Mann, Mass, & Ganzer, 2005). However, MADRS is an observer-based scale, whereas the CSDD is an informant-based scale. This has led the authors to ponder possible differences in symptom focus from observer (clinician) to informant (caregiver). The participants were assessed for dementia using the DSM-IV criteria, and depression diagnosis using the DSM-IV, International Statistical Classification of Diseases version 10 (ICD-10) criteria, and the PDC-dAD. Dementia and its severity were assessed using Brazilian-adapted versions of the Clinical Dementia Rating (CDR) scale, MMSE, and Neuropsychiatric Inventory (NPI). In addition, all patients were screened for depression using the Brazilian version of the MADRS and their appropriate caregiver/s completed the adapted version of the CSDD. The results showed that an MADRS score of 10 and a CSDD score of 13 were the optimal cutoffs to recognize depressive disorder within its sample. Results of the study were not in line with prior studies. Several studies have reported a CSDD cutoff between 5 and 8 for depression in dementia (Alexopoulos, Abrams, Young, & Shamoian, 1988; Barca, Engedal, Laks, & Selbaek, 2010; Barca, Selbaek, Laks, & Engedal, 2008; Schreiner, Hayakawa, Morimoto, & Kakuma, 2003). The authors suggest several possibilities for this, including cultural differences in Latin-American populations and the possibility that carers may have provided false information so as to force admission of the patient. The authors conclude that both the MADRS and the CSDD should be viewed as screening instruments instead of diagnostic measures in samples similar to theirs.

The main aim of one study by Starkstein, Dragovic, Jorge, Brockman, and Robinson (2011) was to validate a set of diagnostic criteria to diagnose depression in AD. Participants recruited were 971 AD patients with very mild, mild, moderate, or severe dementia. All participants attended the Dementia Clinic of a tertiary care center in Buenos Aires, Argentina between January 1993 and October 2002. Initial psychiatric examination was conducted using the Structured Clinical Interview for DSM-IV (SCID). Sad mood was rated according to the DSM-IV definition ("sad mood present most of the day, nearly every day, and for more than 2 weeks"). Other applied examinations were the MMSE, CDR, Hamilton Depression Scale (HAM-D), Hamilton Anxiety Scale, Apathy Scale, and the Irritability Scale. Latent Class Analysis, a cluster procedure, was used to determine the latent structure of patients with AD and to categorize them according to their SCID responses, as well as other psychiatric instruments. The best-fitting model produced by the Latent Class Analysis was a three-class model, with groups defined by their psychiatric symptom profile. Class 3, comprised of 21% of participants, illustrated a high frequency of all nine DSM-IV criteria for major depression. Class 2 comprised 39% of the sample, the frequency of depressive symptoms much broader, ranging from 4% to 55%, equating with minor depression. Class 1, comprised of 40% of participants, displayed a very low frequency of depressive symptoms. The between-group differences for all nine DSM-IV criteria for major depression were statistically significant. 96% of the Class 3 participants met unmodified DSM-IV criteria for major depression, when compared with none in the no-depression cluster. The study also found that anxiety and apathy are significant predictors of depression in AD, while irritability is not. Finally, it was found that participants with minor depression had significantly higher frequency of apathy than patients without depression. This suggests that minor depression in AD could be a diverse condition that includes patients with true depression, and those nondepressed patients with apathy. Further, a significant association between more severe depression and more advanced dementia stages was found. By comparing the diagnosis of major depression using the National Institute of Mental Health-depression in AD (NIMH-dAD) to DSM-IV, 38% of the sample met the modified NIMH-dAD criteria, compared with 27% using DSM-IV criteria. Starkstein et al. (2011) suggest they may have underestimated the frequency of depression based on the NIMH-dAD given that they used the time criteria for frequency of symptoms from the DSM-IV criteria across the entire sample. The DSM-IV time criteria are more stringent than the NIMH-dAD. The authors conclude that their study validates the DSM-IV criteria for major depression for use in AD, arguing it provides the optimal number of symptoms to diagnose major depression.

Vilalta-Franch et al. (2006) argue that depression manifests differently in older depressed patients when compared to their younger counterparts, such that older patients often deny they feel sad and instead report a lack of feeling or emotion, or loss of pleasure and interest in activities. This study aimed to determine the difference in prevalence rates of depression in AD patients through a cross-sectional study, comparing specific diagnostic criteria with other more general criteria. It also aimed to determine the clinical characteristic of patients with depression and without depression according to each diagnostic criteria used. Data in this study were obtained from 491 patients with probable AD who completed the baseline visit of a prior study, the Evolution of Dementia of the Alzheimer-type and Caregiver burden (EDAC) between May 1998 and May 2003. The EDAC was a naturalistic study that took place over a 24-month period. At baseline the Cambridge Examination for Mental Disorder of the Elderly (CAMDEX) and the NPI were used. CAMDEX is a standardized, structured interview and examination for diagnosing common mental disorders in later life with special reference to dementias. It includes an interview with the patient, a cognitive examination, an interview with a relative or caregiver, and multiple scales [Blessed Dementia Rating Scale (BDRS), Ischemia Score, MMSE]. CAMDEX grants clinical diagnosis based on DSM-IV and ICD-10 criteria, primary and secondary psychiatric diagnoses, clinical estimate of severity of dementia, clinical estimate of severity of depressive symptoms (CAMDEX Depression Diagnostic Scale), and other medical diagnoses. The NPI is a rating scale for behavioral and psychological symptoms of dementia administered to a caregiver or family member. The study used the ICD-10 and DSM-IV for diagnosing episodes of major depression, the CAMDEX criteria for diagnosing depression, the PDC-dAD and a positive response to the screening question "Does the patient seem sad or depressed? Does he or she say that he or she feels sad or depressed" within the NPI depression subscale. This equates to five conceptualizations of depression in increasing order of diagnostic rigidity. The results of the study found the prevalence of depression in AD according to each criteria as follows: ICD-10, 4.9%; CAMDEX, 9.8%; DSM-IV, 13.4%; PDC-dAC, 27.4%; NPI depression subscale, 43.7%. Percentage of agreement as measured by the kappa index was low to moderate between the classification systems. No pair of criteria reached a 50% agreement rate. Loss of confidence or self-esteem and irritability were the items that most contributed to the diagnostic disagreement. To conclude, the results of this study indicate the significant variability in prevalence rates of depression in AD depends on which diagnostic criteria is applied. Concordance is low between the measures and the use of specific depression diagnostic criteria increases the prevalence rate.

Reports of the prevalence of depression in dementia vary widely, most likely due to the variance in severity of cognitive impairment and the broad range of scales used for the detection of depression (Muller-Thomsen et al., 2005). Due to the similarity in symptomatology of both depression and dementia, differentiating between the two can be difficult. However, the treatment of depression in dementia is reported to be successful by Lyketsos and Olin (2002). Muller-Thomsen et al. (2005) assessed the prevalence of depression in relation to stage of AD using four different depression scales. The secondary goal was to test the validity of each of these scales by calculating the internal consistency and correlation between them. Participants were 316 patients diagnosed with probable AD according to the National Institute of Neurological and Communicative Disorders and Stroke – Alzheimer's Disease and Related Disorders Association (NINCDS-ADRDA) criteria. Patients had been admitted to the Memory Clinic of the Department of Psychiatry of the University Hospital of Hamburg between 1995 and 2001. The severity of cognitive impairment was determined using the MMSE. Participants were divided into two groups—mildly demented (MMSE score of 18–30) and moderately severely demented (MMSE score of 0–17). To assess depression, the following four scales were used: the GDS—self-rated, the MADRS—assessed by neuropsychologist, the NOSGER—caregiver-rated, and the CSDD—physician-rated. According to the GDS, the prevalence of depression in AD was about 35% when a cutoff of 6 was used for the detection of depression. Muller-Thomsen et al. (2005) found a marked decrease of internal consistency in the moderate to severe AD group, indicating that severity of cognitive decline may affect the results of the GDS. Muller-Thomsen et al. (2005) concluded that the GDS is not an adequate tool for detecting depression in AD, especially in the later stages of the disease. This supports prior findings by Burke, Roccaforte, and Wengel (1991) and Zarb (1996). The results of the caregiver-rated NOSGER mood scale reported markedly higher depression rates than the other scales. However, in accordance with the other scales, the depression rate was higher in the moderate to severe AD (68%) than in mild AD (53%). Three possibilities are presented by the authors to understand the comparatively high prevalence rates reported by the NOSGER mood. First, it is possible the carers are projecting their own personal burden, which in turn influences their answers. Second, the cutoff of the scale is too low. Finally, using subscales of the NOSGER is not meaningful, as mentioned by Muller-Thomsen et al. (2005). The internal consistency of NOSGER mood was poorest of all scales assessed and there was poor correlation with the results of the other scales. Both the MADRS and the CSDD classified approximately 40% of the moderate to severe AD patients as depressed, and 30% of the patients in the mild AD group. Internal consistency was good for both scales and a significant positive correlation between the two assessments was observed. The authors suggest this could be due to the similarity in some items across the two scales. In conclusion, one consistent finding was that depression was more prevalent in the moderate-severe AD group than the mild AD group. MDRS and CSDD are recommended as useful tools for detecting depression and describing depressive symptoms in an out-patient AD population.

The Hamilton Rating Scale for Depression (HRSD) is a commonly used instrument both for clinical and research purposes. Naarding, Leentjens, van Kooten, and Verhey (2002) assessed whether disease-specific cut-off scores should be applied to the scale for various neurological disorders including stroke, AD, and Parkinson's Disease. The authors compared the concurrent validity of the HAM-D in relation to DSM-IV criteria for MDD. The AD group comprised of 274 patients diagnosed with dementia according to the DSM-IV and NINCDS-ADRDA criteria.

The HAM-D was used as a symptom checklist to diagnose MDD according to DSM-IV criteria. The cognitive status of 239 patients was also assessed using the MMSE. The prevalence of MDD was 22.6% in the AD group. In order to determine whether HAM-D could be used as a predictive test, positive predictive values (PPV) and negative predictive values (NPV) were calculated for different cut-off scores in the central range of the scale. Optimal cutoff for diagnostic purposes was found in the AD group at 13/14 for the high PPV and 6/7 for the NPV. Naarding et al. (2002) argued that the psychomotor and autonomic symptoms measured in the HAM-D theoretically could coincide with DSM criteria for MDD thus making diagnosis of MDD more difficult in patients with AD. However, the concurrent validity of the HAM-D with the DSM-IV criteria for MDD was high and this coincidence should therefore not be of clinical concern. In conclusion, optimal performance of the scale requires the use of disease-specific cut-off points for screening, diagnostic, and dichotomization purposes.

It has been previously thought that symptoms of depression in AD differ from those seen in the elderly with normal cognition (Engedal, Barca, Laks, & Selbaek, 2011). It was believed that AD patients tended to have more motivational symptoms and delusions and fewer common core symptoms of depression such as sadness, sleep disturbances, and appetite loss (Olin, Katz, Meyers, Schneider, & Lebowitz, 2002; Olin, Schneider et al., 2002). However, more recent studies have not confirmed these findings, instead reporting that the typical symptoms of depression are also present in dementia patients, especially sadness (Ballard et al., 2000; Chemerinski, Petracca, Sabe, Kremer, & Starkstein, 2001; Starkstein, Mizrahi, & Garau, 2005). As a result of this debate, a group of clinical experts in this field proposed a new set of diagnostic criteria for depression in AD: the PDC-dAD. The few studies that have implemented the new criteria have found that its use leads to higher prevalence rates of depression in AD being reported than those found when using the DSM or ICD criteria. Although there are several studies of the commonly used depression measures in dementia patients such as the HRSD (e.g., Brodaty & Luscombe, 1996), there are also other scales such as the PDC-dAD used specifically for dementia of the Alzheimer's type patients (Sepehry et al., 2017). Engedal et al. (2011) aimed to further investigate the usefulness of the PDC-dAD and to examine which symptoms of depression according to the PDC-dAD compared to those of the DSM-IV-TR and ICD-10 criteria in the same AD patients. Specifically, the identification of symptoms using the PDC-dAD measure was also compared to the dementia depression diagnosis of the PDC-dAD, the Diagnostic and Statistical Manual of Mental Disorders-Fourth Edition in text version using the Major Depression criteria (DSM-IV-TR) and, the ICD-10 measures to compare its effectiveness in identifying and diagnosing depression in AD patients. The study consisted of 112 AD patients, 55 from nursing homes and 57 from geriatric psychiatric hospitals. AD diagnosis was made according to ICD-10 criteria for research. Patients were assessed with the CSDD for depression by trained nurses. ADL was assessed using the Physical Self-maintenance Scale. Cognitive function and stage of dementia were assessed using the MMSE and the CDR scale. Within a week of the CSDD being administered, experienced geriatric psychiatrists (blind to the results of the CSDD) recorded the presence of depressive symptoms as defined by PDC-dAD criteria, the research criteria of the ICD-10, and the DSM-IV-TR (Major Depression) criteria. As was anticipated, the PDC-dAD criteria captured more patients with a diagnosis of depression than did the DSM and ICD criteria. Specifically, the results reported prevalence rates of depression in AD at the following levels: PCD-dAD—53.6%; ICD-10—47.3%; DSM-IV-TR—34.8%. The PDC-dAD criteria had poor results in a Hosmer-Lemeshow goodness of fit test, making the model less valid than the ICD-10 and DSM-IV-TR. This was the case in other studies (Teng et al., 2008; Vilalta-Franch et al., 2006). Reasons for this are proposed to be that only three symptoms are required to make the diagnosis and that the symptoms do not need to be present most of the day, nearly every day. Delusions are also assessed by the PDC-dAD and as most of the patients had moderate to severe AD, delusions are likely to be more frequent. Agreement between the three criteria was high. The pattern of symptoms seen using the DSM and ICD criteria was similar to that seen in a depressed patient without dementia. The six most significant symptoms when comparing the depressed to the nondepressed patients were anxiety, sadness, multiple physical complaints, suicidal thoughts, pessimism, and poor self-esteem. Delusions and agitation were important in the diagnosis according to the ICD and PDC-dAD but not for the DSM. In conclusion, it was found that the most prominent and significant symptoms of depression in patients with AD are the same symptoms considered to be the core symptoms of depression in the elderly that do not suffer dementia. Further, it was confirmed that more patients receive a diagnosis of depression using the PDC-dAD criteria than using the DSM or ICD criteria.

To explore the ability to utilize the NIMH-dAD, Teng et al. (2008) assessed the amount of remissions in depression patients using the NIMH-dAD measurement of depression in dementia in comparison to the diagnoses of depression using other scales of measurement. They recruited participants from AD Research Centers running a descriptive longitudinal study of depression in AD totaled 101 participants who met the DSM-IV criteria for major or minor depression. Participants were assessed at two intervals of the study, at baseline and after 3 months. The assessment at both instances measured depression using several scales such as the NIMH-dAD, DSM-IV, CSDD, NPI-Q (questionnaire version). The DSM-IV criterion used was that of major or minor depression, which was also supported using the Structure Clinical Interview for DSM-IV Axis I Disorders. Other supporting sources of information gathering to support the diagnosis for some measures (e.g., NIMH-dAD) were clinician impressions after interviews with participants and their caregiver. The administration of these assessments and scale measures after 3 months allowed the evaluation of remission of participants. Teng et al. (2008) found that the NIMH-dAD measure identified more AD patients as depressed in comparison to other measures of depression did, thus suggesting the NIMH-dAD scale as more effectively distinguishing depressive symptoms. Although the NIMH-dAD scale identifies and distinguishes patients' depressive symptoms more than other measures, it is of close standard to that of the DSM-IV in diagnosing minor and major depression as they share very similar characteristics such as only needing the presence of few depressive symptoms (three for the NIMH-dAD and five for the DSM-IV) to appear for less restricted amounts of time in order to meet depression criteria. The three symptoms of the NIMH-dAD derived from the DSM-IV are psychomotor changes, fatigue, and a sense of guilt/worthlessness. In addition to these three symptoms, the DSM-IV also includes the symptoms of social isolation/withdrawal and irritability. Although there is an advantage of being precautious and taking individuals' depressive symptoms seriously, there may be a disadvantage of low specificity where individuals may be diagnosed with depression when they are not actually depressed. Such interpretation is understood to be supported by the finding of remission rates of participants being 51% at the 3-month follow-up of the study. Similar remission rates were also found for the other measures of depression and are suggested as attributable to participants undergoing treatment for depression, which may be a confounding or contributing factor to results of the study. In conclusion, it is of great importance to consider the symptoms associated with depression because although they may not be present or reflective of all dementia patients (Teng et al., 2008), there is a variability among the depressive symptoms patients experience and, being more specific rather than over diagnosing depression in patients seems important, especially if therapeutic or drug interventions are later to be implemented.

To evaluate the effectiveness of measures of depression in dementia patients, Gilley and Wilson (1997) used the GDS and the HRSD to diagnose major depression. The study also evaluates the effect of cognitive impairment of the GDS by having two participant groups: cognitively impaired AD patients and cognitively intact participants. Overall, a total of 715 AD patients made up the cognitively impaired group and 93 participants made up the cognitively intact group. All participants of the study underwent depression and cognitive measures. To rate depression, the GDS was used, measuring affective/mood states of participants over the past week along with the criterion measure of an interview with an informant (family member) who has been most consistently in contact with the participant over the past 4 weeks. Also assessing the depressive state of participants, a diagnosis of MDD was done using the Structured Clinical Interview for DSM-III (SCID) in which SCID and HRSD were embedded in the interview as well to rate depression severity stage. Evaluating participants' cognition in exploration of whether cognitive impairment affects the results of the GDS, measures administered to participants included the MMSE to assess global level of cognitive impairment, the Logical Memory and Figural Memory subtests of the Wechsler Memory Scale—Revised to asses memory as well as the subtests of the Multilingual Aphasia Examination (visual naming scale and word fluency with the Controlled Oral Word Association scale) to assess language functioning. To further evaluate whether cognitive impairment had an impact on measures of the GDS, statistical analyses were conducted (e.g., logistic regression, product-moment correlation). Results of the Gilley and Wilson (1997) study included a poor correlation between the GDS measure scores and the criteria of the SCID and HRSD in the AD patient group when compared to the cognitively intact participant group. Moreover, in participants of the AD group, their level of cognition did affect the scores on the GDS, decreasing the validity of the GDS measure used in AD patient groups. Interpreting such findings, it could be suggested that in AD patients using the GDS measure can lead to high false-negative rates in which AD patients may have depression (or symptoms of depression), but their scores of the measure itself indicate that they do not have depression. In disagreement with other studies (e.g., Sheehan, 2012) where the use of the GDS is suggested to be validated in the use of mild stage dementia, it could be concluded that more research should be done in investigating the validity of GDS in mild dementia patients such as the CSDD (e.g., Korner et al., 2006).

Korner et al. (2006) investigated the use of measures of depression in elderly and dementia patients. The study involved evaluating the use of the GDS, CSDD, and HDRS in a total of 145 elderly participants aged 65 or older (73 depressed, 36 depressed and demented, 36 not depressed with 11 of these being demented only). In cases where participants were demented, dementia severity was not specified. Korner et al. (2006) also reviewed the sensitivity and specificity of the GDS and CSDD measures specifically as well as the inter-observer reliability of the CSDD in particular. Doing so, the study employed the method of evaluating participants' depression and dementia using the ICD and Clinical Global Impression measure where interviews took place with participants. To further explore and measure depression in the participant groups, the GDS, CSDD, and HDRS were used; while exploring participants' cognitive impairment, the MMSE was used. Findings of the study included the GDS, CSDD, and HDRS being good measures for depression in general; however, in demented or cognitively impaired participant groups as well as the nondemented participant groups specifically, the CSDD was more appropriate than the GDS in assessing depression, although the CSDD and GDS were closely related to the HDRS scale. In terms of scale content, the two measures of CSDD and HDRS had a strong relationship in which four scale items assessed depressive symptoms (specifically mood, diminished interests, low energy, and suicidal thoughts) as well as some sleep and psychotic symptoms relating to depression. Interpreting these findings in relation to the effectiveness of measures assessing depression in dementia, it could be argued that the GDS is not appropriate in demented patients and that the CSDD also competes with appropriateness and suitability with the HDRS for dementia patients when screening for depression. However, when it comes to also being suitable for measuring depression in dementia patients, it is suggested by other studies that the HDRS is not effective, relating to the symptoms of depression such as the inability to maintain conversation as the scale takes approximately 20 − 30 minutes to complete (Sheehan, 2012), thus leaving the CSDD as the most effective and appropriate scale of measuring depression in dementia. In line with the idea of scale length being appropriate for use in dementia patients, Korner et al. (2006) also advise that the length of measures is a vital characteristic of measure effectiveness, especially in old aged individuals who may also be demented. It is advised that to reach high levels of sensitivity and specificity, measures of depression in both demented and nondemented elderly populations, length of scales should be of four items.

Vida, Des Rosiers, Carrier, and Gauthier (1994) compared the ability of the CSDD and HDRS measures in detecting depression in AD patients and evaluated the appropriateness of the scales using specificity and sensitivity using nonparametric statistical analysis of the receiver operating characteristic (ROC) of each scale. Doing so, the study also evaluated correlates of the CSDD and HDRS scales with the Research Diagnostic Criteria (RDC) of major depression. 10 participants were involved in the study and took part in an interview (along with their carers) to gather data regarding demographics, social, and medical information. Moreover, participants' cognitive and depressive mood states were rated using several measures: the Global Deterioration Scale (GDS), MMSE, HDRS, and CSDD. Following these measures, participants took part in another interview where researchers used the RDC to test (and rate) major depression in terms of AD mild to moderate severity stages. Findings of the study included moderate to high correlation between both scales (CSDD and HDRS) with the RDC. Although a good relationship was found between both scales with the RDC, the CSDD slightly correlated more than the HDRS with the RDC, suggesting a stronger relation in terms of measuring depression in AD patients. Moreover, in analysis of specificity and sensitivity, both the CSDD and HDRS were indicated to be effective measures in differentiating the diagnostic criteria for mild to moderate AD patients to meet depression criteria. Such differentiation is important since depression going undiagnosed in AD patients in their early stages of dementia could lead to severe consequences such as hopelessness and suicide ideation (Harwood & Sultzer, 2002; Sabodash, Mendez, Fong, & Hsiao, 2013). Therefore, the CSDD may be more effective than the HDRS in identifying depression in dementia patients, which is in line with other literature (Korner et al., 2006) that suggest the CSDD as having good content validity and functionality since it is a short scale that can be administered with more ease in dementia patients with low concentration. Thus it is clear that the CSDD is more effective than the HDRS as it has characteristics outlined and highlighted to be desirable in measures of depression in dementia such as good sensitivity and specificity.

Lam et al. (2004) evaluated the effectiveness of the GDS, Even Briefer Assessment Scale for Depression (EBAS DEP), and the Single Question as well as the CSDD in measuring and diagnosing depression in dementia. A total of 88 dementia patients aged 60 years or over were involved in the study (66 without depression; 22 with depression). Participants were grouped according to dementia severity stages (mild and moderate to severe) for comparison of how well the scales diagnosed depression along the severity stages of dementia. The four scales of interest in the study were used to assess participants' depression by a research psychologist blind to whether participants had depression or not upon recruitment. All scale responses were obtained from participants' caregivers who knew them best in regards to consistent and most contact and knowledge of their ADL. However, the Single Question was directly administered to patients only. To measure the effectiveness of the scales and the influence of gender, age, and dementia severity stage between the depressed and nondepressed dementia participant groups, and if the Single Question could diagnose depression, statistical analyses were employed such as Chi-Square and Mann – Whitney U tests. Another method of analysis used to compare the effectiveness of the GDS, EBAS DEP, and Cornell scales was the ROC. This method was used to assess the sensitivity and specificities of the measures. Using cross-tabulation, another statistical method, the Single Question measure's specificity and sensitivity was also evaluated. Findings of the study show that as dementia gets worse in terms of cognitive impairment, depression diagnosis becomes more difficult possibly due to memory loss and the more specific selection of scales that are not lengthy to suit patients' symptoms of lack of attention and concentration in late dementia (Brodaty & Luscombe, 1996; Korner et al., 2006; Sheehan, 2012). Thus it is suggested that in moderate to severe stages of dementia, the cut-off scores for depression should be higher. Furthermore, the utilization of the Single Questionnaire appeared to be more powerful in diagnosing depression than the other scales (GDS, EBAS DEP, CSDD, and Single Question), especially in the later stages of dementia. Following the Single Question measure being the most appropriate measure, the CSDD was next most effective in diagnosing depression in participants who had dementia. As such, Lam et al. (2004) suggest that when selecting a measure to diagnose depression in dementia patients, the Single Question scale should be used, followed by the CSDD if necessary since the CSDD had the most sensitivity and specificity in diagnosing depression, even when regarding dementia severity stages.

Although there are many scales used to assess depression, they tend to measure depression severity more than identifying if individuals actually have depression or not. There are also different scales that should be preferably used for patients with cognitive decline (dementia) and others without cognitive decline. That is, the MADRS is preferred for use in measuring depression in cognitive normal elderly and the CSDD in elderly people with both dementia and normal cognition (Engedal et al., 2012). Therefore, the aim of the current study was to evaluate the validity of the MADRS in identifying depression in cognitively normal individuals who had been diagnosed with depression by psychiatrists using the DSM-IV and compare the MADRS validity with the CSDD. 104 participants over 65 years of age who did not have signs of dementia were involved in the study (101 women, 39 men) and took part in cognitive evaluation using the MMSE. Moreover, assessment of depression included participating in an interview as part of the MADRS and its scale as well as caregivers and the patients being interviewed as part of the CSDD measure. In addition to using the two-scale measures of the MADRS and CSDD, depression was diagnosed using the DSM-IV for all participants whereas only 70 participants were diagnosed for depression using the ICD-10. Results of the study indicate that the MADRS is suitable for diagnosis of depression in nondemented persons in which it has good sensitivity and specificity. In comparison to the DSM-IV, the ICD-10 was found to be less strict in the diagnosis of depression (in terms of criteria for major depression). More specifically comparing the ICD-10 with the MADRS, the ICD-10 was more accurate in identifying depression when not comparing these measures with the DSM-IV. Such findings may be related to the MADRS not being effective in identifying depression in dementia patients when compared to the ICD-10, and the ICD-10 also having a less strict criterion in identifying and diagnosing depression than the DSM-IV. Moreover, when regarding the CSDD, in comparison, the MADRS is suggested to be a more useful measure in elderly nondemented participants. Therefore, it could be concluded that not only is the MADRS a more useful measure for depression in nondemented patients but, it is also better at discriminating which participants have depression and which do not when compared to the CSDD.

Muller-Thomsen et al. (2005) explored the presence of depression along AD stages measured using four different depression scales namely the GDS, MADRS, CSDD, and the NOSGER. The study involved exploring 316 cases of patients with probable AD, which was split into two groups: mild AD and mild to severe AD according to measures of the MMSE to assess whether cognitive impairment influenced measures of depression. The following scales were administered by individual professionals (e.g., physician, neuropsychologist) or caregivers to assess participants' depression: GDS, MADRS, CSDD, and NOSGER, unless too cognitively impaired to complete items of scales. Following statistical analysis using SPSS, findings overall implied that there was less depression in participants in the moderate to severe AD groups (15.7%) in comparison to the mild AD group (26.8%). Specifically regarding the diagnosis of depression in the two groups using the four separate scales, 35% of participants were found to be depressed according to the GDS measure. It was noted by authors that the internal consistency of participants' diagnosis of depression within the moderate to severe AD group was low, meaning that the GDS may not be a tool of strength to use when assessing for depression in the later stages of AD. Moreover, using the NOSGER mood scale, depression identification among the participants was quite high in comparison to the other three scales of investigation, especially in participants of the moderate to severe AD participant group. However, such finding is suggested to be influenced by caregiver burden among other reasons including a lack of internal consistency, one poorer than the rest of the other scales of interest in the study. Identified with adequate internal consistency and a positive correlation, the MADRS and CSDD found 40% of participants of the moderate to severe AD group and 30% of the mild AD group to have depression. In general, the finding that depression was detected more in the moderate to severe groups suggests that the neurological progression of AD may influence the appearance or development of depressive symptoms as suggested by the authors. In conclusion, what can further be presented is the idea that the MADRS and CSDD appear to be the most suitable of the four scales examined in the investigation of detecting depression along the path of the disease progression of AD, which is supported by the scales' internal consistency and correlation. Although both scales are suggested to be good at detecting depression, some administers prefer one over the over for reasons such as the CSDD being easier to use as its items are worded more subjectively in comparison to the MADRS in which its items were more objective and easier to understand and apply to the patients (Leontjevas, van Hooren, & Mulders, 2009). Furthermore, the suitability of using the MADRS and CSDD as measures in effectively detecting depression in dementia is also supported by another study finding the scales to also have good internal consistency, and to be good at distinguishing depressed from nondepressed dementia participants (Leontjevas et al., 2009).

Linking the symptoms of depression in dementia patients with specific symptom characteristics of measures of depression, Vilalta-Franch et al. (2006) investigated the prevalence of depression diagnosis using various scales of depression among a population of 491 patients with probable

AD. Following cognitive and functional assessment measured using scales such as the Cambridge Cognitive Examination, MMSE, and BDRS, at baseline of the study, participants were completed the CAMDEX and the NPI for clinical diagnosis of dementia along with the NPI to rate behavioral and neuropsychological symptoms of the disease. To assess for depression, the patients completed the following measures: ICD-10, DSM-IV for episodes of major depression, the CAMDEX (depression), the PDC-dAD, and a positive response to a screening question using the NPI depression subscale. Findings of the study did indeed show variability in detecting depression among the different scales. The ICD-10 identified 4.9% of the population as depressed, the PDC-dAD 27.4%, the DSM-IV 13.4%, the CAMDEX 9.8%, and the NPI depression subscale 43.7%, which is suggested to be overrated by caregiver overemphasis of patient symptom severity. Comparing the prevalence rates detected by the scales revealed that the PDC-dAD more specifically identified symptoms of depression that the ICD-10, CAMDEX, and DSM-IV do not consider when diagnosing depression in dementia patients. Findings of the study results suggesting the PDC-dAD scale to be most effective in identifying depression in dementia patients is supported by the scale having the most agreement with the other scales used in the study when compared, specifically when the DSM-IV, CAMDEX, and NPI have low levels of agreement between each other, and the NPI also having a 15% false-negative rate in detecting depression in which it screens individuals as not being depressed when they actually do have depression.

Discussion and conclusion

In conclusion, our review here shows that depression has a widespread effect on several domains (both psychological and physical) in older adults, individuals with mild cognitive impairment, and dementia patients. Importantly, the relationship between dementia and depression is bidirectional, such that a history of depression was found to increase the chance of developing AD as well as dementia symptoms (cognitive decline, executive dysfunction, among other cognitive impairment) do all impact mood and lead to depression in AD. Importantly, the treatment of depression in AD (as well as individuals with mild cognitive impairment and subjective cognitive impairment) does also ameliorate cognitive decline. While there have been other reviews on the relationship between depression and dementia (Table 1.2, also see Korner et al., 2006; Monastero,

Table 1.2 Some prior representative reviews on the relationship between dementia and depression.

Scales/domain	Comments
	Sheehan (2012) **Reviews tests of dementia severity and assists suggestive treatment facilities.**
Cornell Scale for patient and carers.	Pros. It has been validated for patients with and without dementia, and is considered by some as the gold standard for diagnosing depressive symptoms in dementia. Authors: Both patient and carer can answer questions; so if the patient is unable to answer the carer can, making it very useful.
Montgomery-Asberg Depression Rating Scale	Assesses psychological symptoms. Pros. Limited usefulness past mild dementia. Authors: It is good to use in intervention research. We think it will be useful to evaluate the effectiveness of interventions by (assumingly) comparing depression (prior intervention) and current depression (postintervention)—that is, history versus current depression.
Hamilton Depression Rating Scale	Pros: Most commonly used for depression assessment. Questions and interview. Cons: Unlikely to be used in people with dementia. Authors: Unlikely to be used in dementia patients. As symptoms of dementia include difficulty to keep up or follow conversations, we think it might be difficult to administer this scale as it includes a semistructured interview that takes around 20–30 min.
Geriatric Depression Scale	Pros: Most commonly used assessment of depressed mood among older people. Self-rating or rated by assessor. Reliable in elderly in institutional care. Cons: Sensitive to change. Validated for mild dementia, but not moderate to severe dementia—completion rates low due to difficulty comprehending questions. Due to pros and cons, we think it might be good to use in mild dementia only.

(*Continued*)

Table 1.2 (Continued)

Scales/domain	Comments
Hospital Anxiety and Depression Scale	Pros: Screening test for depression and anxiety aimed at hospitalized patients. Easy to use and accurate at detecting depression. Cons: little practical use for older patients with significant cognitive impairment as response to items is subjective. We think it is not as good as other scales to be used with dementia patients past mild stages because limited or no insight of their symptoms may affect their responses to items.
Kirkham et al. (2016) **Addresses the accuracy of depression rating scales for diagnosing depression in dementia in AD, vascular dementia, and dementia with Lewy bodies.**	
Cornell Scale for Depression in Dementia	Specific to depression in dementia. Authors: ≥ 9 screen positive.
Dementia Mood Assessment Scale	Caregiver reports based on clinician interview. Contains both mood and cognitive assessments.
Nurses' Observation Scale for Geriatric Patients	Caregiver report. Authors: ≥ 10 suggestive of depression.
Geriatric Depression Scale	Self-report. $\geq$ 6, $\geq$ 8, or $\geq$ 9 is indicative of depression.
Hamilton Depression Rating Scale	Clinician interview. ≥ 8 considered positive for depression.
Montgomery-Åsberg Depression Scale	Clinician interview. ≥ 13 screen positive for depression.
Beck Depression Index	Self-report. ≥ 13 screen positive for depression.
Patient Health Questionnaire	Self-report. $\geq 8 - 11$ suggestive of depression.
Thrope (2009) **Discusses symptoms in depression and dementia**	
Cognitive and functional decline	Excessive preoccupation with deficits. Lack of concern or denial about symptoms. Authors: It is very difficult to make a firm diagnosis of depression in dementia, especially when advanced.
Mood	Mostly sad and stimulation/support does not help much. Mostly normal. Depends on circumstances and fluctuates. If low mood, support helps.

(Continued)

Table 1.2 (Continued)

Scales/domain	Comments
Interest/initiative	Loss of interest and pleasure over few months, includes feeling sad, guilt, self-harm, hopelessness. Apathy over long periods of time but, does not involve sadness.
Eating behavior and weight loss	Changes in appetite over weeks leading to increase/decrease in weight. Gradual weight loss over months to years. Increase in height due to decreased activity, medications.
Sleep	Disturbed sleep cycle (over months to years) with brain changes. Night time waking and daytime sleeping. Increase or decrease in sleep over a few weeks.
Psychomotor agitation	Happens over weeks, worse in mornings but occurs during the day too. Includes nihilistic statements (e.g., life is meaningless) or excessive guilt. Happens over months or years, worse when it comes to the end of the day and later in disease progression. Patients start seeking people or places from their earlier life experiences.
Psychomotor retardation	Occurs in severe depression over periods of weeks. Occurs in mild to moderate dementia and more often in later stages. Can be mimicked by Parkinson's (facial masking, slow motor functioning)/Pick's disease.
Energy	Low energy and complaints of fatigue. Normal energy but, reduced activity due to slow initiation due to slower executive functioning.
Guilt/worthlessness	Common in severe depression. Involves low mood and changes in appetite and sleep. Statements of guilt/worthlessness are common when stressed with patients who still have insight of disease.
Concentration and thinking	Loss of concentration and focus, indecisive and afraid of making mistakes. Things like thinking are affected in later stages of disease.

Mangialasche, Camarda, Ercolani, & Camarda, 2009; Novais & Starkstein, 2015; Wragg & Jeste, 1989), here we attempted to also focus on the several measures used to assess depression in dementia to possibly explain conflicting findings in the literature.

The prevalence of depression in AD varies from one study to another, with several studies reporting rates from 20% to 80%. One reason underlying this discrepancy is the use of many different scales to measure depression (see Table 1.1). Some of these scales include Beck's Depression Inventory, CSDD, DMAS, GDS, HAM-D, Hospital Anxiety and Depression Scale, MADRS, NOSGER Patient Health Questionnaire, NIMH PDC-dAD, CAMDEX, NPI ICD-10 DMAS Alzheimer's Mood Scale, Zung self-rating depression scale, among others. One problem is, most of these scales measure different aspects of depression, thus leading to problems accurately estimating rate of depression in AD.

The relationship between insight (awareness of one's symptoms) and depression is not straightforward, as some studies report the counterintuitive finding that dementia patients with good insight have more depressive symptoms. It is possible that being aware of one's symptoms makes one feel bad about their condition, and thus lead to depression. Further, the relationship between the severity of patients' symptoms and the well-being of their caregivers has been investigated. Some of these studies show that the more severe the dementia is, the lower the QOL of their caregivers. However, future research should also investigate the relationship of how depression in one group (patients or their caregivers) impacts the other group.

References

Alexopoulos, G. S., Abrams, R. C., Young, R. C., & Shamoian, C. A. (1988). Cornell Scale for Depression in Dementia. *Biological Psychiatry*, *23*(3), 271–284.

Ball, S. L., Holland, A. J., Huppert, F. A., Treppner, P., Watson, P., & Hon, J. (2004). The modified CAMDEX informant interview is a valid and reliable tool for use in the diagnosis of dementia in adults with Down's syndrome. *Journal of Intellectual Disability Research: JIDR*, *48*(Pt 6), 611–620. Available from https://doi.org/10.1111/j.1365-2788.2004.00630.x.

Ballard, C., Neill, D., O'Brien, J., McKeith, I. G., Ince, P., & Perry, R. (2000). Anxiety, depression and psychosis in vascular dementia: Prevalence and associations. *Journal of Affective Disorders*, *59*(2), 97–106.

Barca, M. L., Engedal, K., Laks, J., & Selbaek, G. (2010). A 12 months follow-up study of depression among nursing-home patients in Norway. *Journal of Affective Disorders*, *120*(1–3), 141–148. Available from https://doi.org/10.1016/j.jad.2009.04.028.

Barca, M. L., Selbaek, G., Laks, J., & Engedal, K. (2008). The pattern of depressive symptoms and factor analysis of the Cornell Scale among patients in Norwegian nursing

homes. *International Journal of Geriatric Psychiatry, 23*(10), 1058–1065. Available from https://doi.org/10.1002/gps.2033.
Brodaty, H., & Luscombe, G. (1996). Depression in persons with dementia. *International Psychogeriatrics/IPA, 8*(4), 609–622.
Burke, A. D., Goldfarb, D., Bollam, P., & Khokher, S. (2019). Diagnosing and treating depression in patients with Alzheimer's disease. *Neurology Therapy, 8*(2), 325–350. Available from https://doi.org/10.1007/s40120-019-00148-5.
Burke, W. J., Roccaforte, W. H., & Wengel, S. P. (1991). The short form of the Geriatric Depression Scale: A comparison with the 30-item form. *Journal of Geriatric Psychiatry and Neurology, 4*(3), 173–178.
Chemerinski, E., Petracca, G., Sabe, L., Kremer, J., & Starkstein, S. E. (2001). The specificity of depressive symptoms in patients with Alzheimer's disease. *The American Journal of Psychiatry, 158*(1), 68–72. Available from https://doi.org/10.1176/appi.ajp.158.1.68.
Chiu, I., Piguet, O., Diehl-Schmid, J., Riedl, L., Beck, J., Leyhe, T., & Sollberger, M. (2018). Facial emotion recognition performance differentiates between behavioral variant frontotemporal dementia and major depressive disorder. *The Journal of Clinical Psychiatry, 79*(1). Available from https://doi.org/10.4088/JCP.16m11342.
Connor, D. J., Sabbagh, M. N., & Cummings, J. L. (2008). Comment on administration and scoring of the Neuropsychiatric Inventory in clinical trials. *Alzheimer's & Dementia: The Journal of the Alzheimer's Association, 4*(6), 390–394. Available from https://doi.org/10.1016/j.jalz.2008.09.002.
Dias, N. S., Barbosa, I. G., Kuang, W., & Teixeira, A. L. (2020). Depressive disorders in the elderly and dementia: An update. *Dementia & Neuropsychologia, 14*(1), 1–6. Available from https://doi.org/10.1590/1980-57642020dn14-010001.
Engedal, K., Barca, M. L., Laks, J., & Selbaek, G. (2011). Depression in Alzheimer's disease: Specificity of depressive symptoms using three different clinical criteria. *International Journal of Geriatric Psychiatry, 26*(9), 944–951. Available from https://doi.org/10.1002/gps.2631.
Engedal, K., Kvaal, K., Korsnes, M., Barca, M. L., Borza, T., Selbaek, G., & Aakhus, E. (2012). The validity of the Montgomery-Aasberg Depression Rating Scale as a screening tool for depression in later life. *Journal of Affective Disorders, 141*(2 – 3), 227–232. Available from https://doi.org/10.1016/j.jad.2012.02.042.
Gilley, D. W., & Wilson, R. S. (1997). Criterion-related validity of the Geriatric Depression Scale in Alzheimer's disease. *Journal of Clinical and Experimental Neuropsychology, 19*(4), 489–499. Available from https://doi.org/10.1080/01688639708403739.
Gottlieb, G. L., Gur, R. E., & Gur, R. C. (1988). Reliability of psychiatric scales in patients with dementia of the Alzheimer type. *The American Journal of Psychiatry, 145*(7), 857–860. Available from https://doi.org/10.1176/ajp.145.7.857.
Harwood, D. G., & Sultzer, D. L. (2002). "Life is not worth living": Hopelessness in Alzheimer's disease. *Journal of Geriatric Psychiatry and Neurology, 15*(1), 38–43. Available from https://doi.org/10.1177/089198870201500108.
Kirkham, J. G., Takwoingi, Y., Quinn, T. J., Rapoport, M., Lanctôt, K. L., Maxwell, C. J., & Seitz, D. P. (2016). Depression rating scales for detection of major depression in people with dementia (Protocol). *Cochrane Database of Systematic Reviews, 8.*
Knapskog, A. B., Barca, M. L., & Engedal, K. (2013). A comparison of the Cornell Scale for Depression in Dementia and the Montgomery-Aasberg Depression Rating Scale in a memory clinic population. *Dementia and Geriatric Cognitive Disorders, 35*(5–6), 256–265. Available from https://doi.org/10.1159/000348345.
Korner, A., Lauritzen, L., Abelskov, K., Gulmann, N., Marie Brodersen, A., Wedervang-Jensen, T., & Marie Kjeldgaard, K. (2006). The Geriatric Depression Scale and the Cornell Scale for Depression in Dementia. A validity study. *Nordic Journal of Psychiatry, 60*(5), 360–364. Available from https://doi.org/10.1080/08039480600937066.

Kroenke, K., Spitzer, R. L., & Williams, J. B. (2003). The Patient Health Questionnaire-2: Validity of a two-item depression screener. *Medical Care*, *41*(11), 1284–1292. Available from https://doi.org/10.1097/01.MLR.0000093487.78664.3C.

Kroenke, K., Spitzer, R. L., Williams, J. B., & Lowe, B. (2010). The Patient Health Questionnaire Somatic, Anxiety, and Depressive Symptom Scales: A systematic review. *General Hospital Psychiatry*, *32*(4), 345–359. Available from https://doi.org/10.1016/j.genhosppsych.2010.03.006.

Lam, C. K., Lim, P. P., Low, B. L., Ng, L. L., Chiam, P. C., & Sahadevan, S. (2004). Depression in dementia: A comparative and validation study of four brief scales in the elderly Chinese. *International Journal of Geriatric Psychiatry*, *19*(5), 422–428. Available from https://doi.org/10.1002/gps.1098.

Leontjevas, R., van Hooren, S., & Mulders, A. (2009). The Montgomery-Asberg Depression Rating Scale and the Cornell Scale for Depression in Dementia: A validation study with patients exhibiting early-onset dementia. *The American Journal of Geriatric Psychiatry: Official Journal of the American Association for Geriatric Psychiatry*, *17*(1), 56–64. Available from https://doi.org/10.1097/JGP.0b013e31818b4111.

Lyketsos, C. G., & Olin, J. (2002). Depression in Alzheimer's disease: Overview and treatment. *Biological Psychiatry*, *52*(3), 243–252.

Martinelli, J. E., Cecato, J. F., Bartholomeu, D., & Montiel, J. M. (2014). Comparison of the diagnostic accuracy of neuropsychological tests in differentiating Alzheimer's disease from mild cognitive impairment: Can the Montreal Cognitive Assessment be better than the Cambridge Cognitive Examination? *Dementia and Geriatric Cognitive Disorders Extra*, *4*(2), 113–121. Available from https://doi.org/10.1159/000360279.

Monastero, R., Mangialasche, F., Camarda, C., Ercolani, S., & Camarda, R. (2009). A systematic review of neuropsychiatric symptoms in mild cognitive impairment. *Journal of Alzheimer's Disease: JAD*, *18*(1), 11–30. Available from https://doi.org/10.3233/JAD-2009-1120.

Muller-Thomsen, T., Arlt, S., Mann, U., Mass, R., & Ganzer, S. (2005). Detecting depression in Alzheimer's disease: Evaluation of four different scales. *Archives of Clinical Neuropsychology: The Official Journal of the National Academy of Neuropsychologists*, *20*(2), 271–276. Available from https://doi.org/10.1016/j.acn.2004.03.010.

Naarding, P., Leentjens, A. F., van Kooten, F., & Verhey, F. R. (2002). Disease-specific properties of the Rating Scale for Depression in patients with stroke, Alzheimer's dementia, and Parkinson's disease. *The Journal of Neuropsychiatry and Clinical Neurosciences*, *14*(3), 329–334. Available from https://doi.org/10.1176/jnp.14.3.329.

Novais, F., & Starkstein, S. (2015). Phenomenology of depression in Alzheimer's disease. *Journal of Alzheimer's Disease: JAD*, *47*(4), 845–855. Available from https://doi.org/10.3233/JAD-148004.

Olin, J. T., Katz, I. R., Meyers, B. S., Schneider, L. S., & Lebowitz, B. D. (2002). Provisional diagnostic criteria for depression of Alzheimer disease: Rationale and background. *The American Journal of Geriatric Psychiatry: Official Journal of the American Association for Geriatric Psychiatry*, *10*(2), 129–141.

Olin, J. T., Schneider, L. S., Katz, I. R., Meyers, B. S., Alexopoulos, G. S., Breitner, J. C., & Lebowitz, B. D. (2002). Provisional diagnostic criteria for depression of Alzheimer disease. *The American Journal of Geriatric Psychiatry: Official Journal of the American Association for Geriatric Psychiatry*, *10*(2), 125–128.

O'Shea, D. M., Dotson, V. M., Woods, A. J., Porges, E. C., Williamson, J. B., O'Shea, A., & Cohen, R. (2018). Depressive symptom dimensions and their association with hippocampal and entorhinal cortex volumes in community dwelling older adults. *Frontiers in Aging Neuroscience*, *10*, 40. Available from https://doi.org/10.3389/fnagi.2018.00040.

Portugal Mda, G., Coutinho, E. S., Almeida, C., Barca, M. L., Knapskog, A. B., Engedal, K., & Laks, J. (2012). Validation of Montgomery-Asberg Rating Scale and Cornell Scale for Depression in Dementia in Brazilian elderly patients. *International Psychogeriatrics / IPA*, *24*(8), 1291–1298. Available from https://doi.org/10.1017/S1041610211002250.

Reichman, W. E., & Coyne, A. C. (1995). Depressive symptoms in Alzheimer's disease and multi-infarct dementia. *Journal of Geriatric Psychiatry and Neurology*, *8*(2), 96–99. Available from https://doi.org/10.1177/089198879500800203.

Robert, P., Ferris, S., Gauthier, S., Ihl, R., Winblad, B., & Tennigkeit, F. (2010). Review of Alzheimer's disease scales: Is there a need for a new multi-domain scale for therapy evaluation in medical practice? *Alzheimer's Research & Therapy*, *2*(4), 24. Available from https://doi.org/10.1186/alzrt48.

Sabodash, V., Mendez, M. F., Fong, S., & Hsiao, J. J. (2013). Suicidal behavior in dementia: A special risk in semantic dementia. *American Journal of Alzheimer's Disease and Other Dementias*, *28*(6), 592–599. Available from https://doi.org/10.1177/1533317513494447.

Schreiner, A. S., Hayakawa, H., Morimoto, T., & Kakuma, T. (2003). Screening for late life depression: cut-off scores for the Geriatric Depression Scale and the Cornell Scale for Depression in Dementia among Japanese subjects. *International Journal of Geriatric Psychiatry*, *18*(6), 498–505. Available from https://doi.org/10.1002/gps.880.

Sepehry, A. A., Lee, P. E., Hsiung, G. R., Beattie, B. L., Feldman, H. H., & Jacova, C. (2017). The 2002 NIMH Provisional Diagnostic Criteria for Depression of Alzheimer's Disease (PDC-dAD): Gauging their validity over a decade later. *Journal of Alzheimer's Disease: JAD*, *58*(2), 449–462. Available from https://doi.org/10.3233/JAD-161061.

Sheehan, B. (2012). Assessment scales in dementia. *Therapeutic Advances in Neurological Disorders*, *5*(6), 349–358. Available from https://doi.org/10.1177/1756285612455733.

Starkstein, S. E., Dragovic, M., Jorge, R., Brockman, S., & Robinson, R. G. (2011). Diagnostic criteria for depression in Alzheimer disease: A study of symptom patterns using latent class analysis. *The American Journal of Geriatric Psychiatry: Official Journal of the American Association for Geriatric Psychiatry*, *19*(6), 551–558. Available from https://doi.org/10.1097/JGP.0b013e3181ec897f.

Starkstein, S. E., Mizrahi, R., & Garau, L. (2005). Specificity of symptoms of depression in Alzheimer disease: A longitudinal analysis. *The American Journal of Geriatric Psychiatry: Official Journal of the American Association for Geriatric Psychiatry*, *13*(9), 802–807. Available from https://doi.org/10.1176/appi.ajgp.13.9.802.

Teng, E., Ringman, J. M., Ross, L. K., Mulnard, R. A., Dick, M. B., & Bartzokis, G. (2008). Diagnosing depression in Alzheimer disease with the national institute of mental health provisional criteria, Alzheimer's Disease Research Centers of California-Depression in Alzheimer's Disease, I*The American Journal of Geriatric Psychiatry: Official Journal of the American Association for Geriatric Psychiatry*, *16*(6), 469–477. Available from https://doi.org/10.1097/JGP.0b013e318165dbae.

Teresi, J., Abrams, R., Holmes, D., Ramirez, M., & Eimicke, J. (2001). Prevalence of depression and depression recognition in nursing homes. *Social Psychiatry and Psychiatric Epidemiology*, *36*(12), 613–620.

Thorpe, L. (2009). Depression vs. Dementia: How Do We Assess? *The Canadian Review of Alzheimer's Disease and Other Dementias*, *12*(3), 17–21.

Vida, S., Des Rosiers, P., Carrier, L., & Gauthier, S. (1994). Depression in Alzheimer's disease: Receiver operating characteristic analysis of the Cornell Scale for Depression in Dementia and the Hamilton Depression Scale. *Journal of Geriatric Psychiatry and Neurology*, 7(3), 159–162. Available from https://doi.org/10.1177/089198879400700306.

Vilalta-Franch, J., Garre-Olmo, J., Lopez-Pousa, S., Turon-Estrada, A., Lozano-Gallego, M., Hernandez-Ferrandiz, M., & Feijoo-Lorza, R. (2006). Comparison of different clinical diagnostic criteria for depression in Alzheimer disease. *The American Journal of Geriatric Psychiatry: Official Journal of the American Association for Geriatric Psychiatry, 14*(7), 589–597. Available from https://doi.org/10.1097/01.JGP.0000209396.15788.9d.

Wang, J., Li, W., Yue, L., Hong, B., An, N., Li, G., & Xiao, S. (2018). The study of White Matter Hyperintensity (WMH) and factors related to geriatric late-onset depression. *Shanghai Archives of Psychiatry, 30*(1), 12–19. Available from https://doi.org/10.11919/j.issn.1002-0829.217038.

Wragg, R. E., & Jeste, D. V. (1989). Overview of depression and psychosis in Alzheimer's disease. *The American Journal of Psychiatry, 146*(5), 577–587. Available from https://doi.org/10.1176/ajp.146.5.577.

Zarb, J. (1996). Correlates of depression in cognitively impaired hospitalized elderly referred for neuropsychological assessment. *Journal of Clinical and Experimental Neuropsychology, 18*(5), 713–723. Available from https://doi.org/10.1080/01688639608408294.

Zlatar, Z. Z., Muniz, M., Galasko, D., & Salmon, D. P. (2017). Subjective cognitive decline correlates with depression symptoms and not with concurrent objective cognition in a clinic-based sample of older adults. *The Journals of Gerontology. Series B, Psychological Sciences and Social Sciences*. Available from https://doi.org/10.1093/geronb/gbw207.

Zung, W. W. (1965). A self-rating depression scale. *Archives of General Psychiatry, 12*, 63–70.

Zung, W. W., Richards, C. B., & Short, M. J. (1965). Self-rating depression scale in an outpatient clinic. Further validation of the SDS. *Archives of General Psychiatry, 13*(6), 508–515.

CHAPTER 2

The nature of depression in dementia: a narrative review

Ahmed A. Moustafa[1,2,3], Phoebe Bailey[1,2], Wafa Jaroudi[4], Lily Bilson[1], Mohamad El Haj[5,6,7] and Eid Abo hamza[8]
[1]School of Psychology, Western Sydney University, Sydney, NSW, Australia
[2]MARCS Institute for Brain, Behaviour, and Development, Western Sydney University, Sydney, NSW, Australia
[3]Department of Human Anatomy and Physiology, The Faculty of Health Sciences University of Johannesburg, Johannesburg, South Africa
[4]School of Social Sciences and Psychology, Western Sydney University, Sydney, NSW, Australia
[5]Laboratoire de Psychologie des Pays de la Loire (LPPL-EA 4638), Nantes Université, Univ Angers, Nantes, France
[6]Unité de Gériatrie, Centre Hospitalier de Tourcoing, Tourcoing, France
[7]Institut Universitaire de France, Paris, France
[8]Faculty of Education, Department of Mental Health, Tanta University, Tanta, Egypt

Introduction: the diagnosis, prevalence, and impact of depression in dementia

Depression is comorbid with various types of dementias (Amore, Tagariello, Laterza, & Savoia, 2007; Baruch, Burgess, Pillai, & Allan, 2019; Kuring, Mathias, & Ward, 2018). Depression is reported to be one of the highest mental health problems occurring in the elderly population and a common symptom of individuals with dementia (Australian Bureau of Statistics, 2015). In most Alzheimer's disease (AD) patients, depressive symptomatology presents as mild to moderate. Severe forms of depression are less prevalent; 5%–30% of AD patients may be diagnosed with a major depressive episode (Reichman & Coyne, 1995). Depressive symptoms in dementia include apathy, slow movement, dysthymia, or minor and atypical depressive syndromes (Teri & Wagner, 1992). Several studies have shown that depression in AD represents a relapse of preexisting depression in less than 10% of AD patients (Jaroudi et al., 2017). However, for most AD patients, the symptoms of depression are developing for the first time.

Depression in dementia is important to investigate as it negatively impacts several aspects of patients' lives. For example, studies have found that depression can increase hippocampal plaques and tangles in AD patients (Rapp et al., 2006). Depression in AD is often comorbid with apathy (Lyketsos et al., 2000) and anxiety (Porter et al., 2003). Benoit et al. evaluated the

Alzheimer's Disease
DOI: https://doi.org/10.1016/B978-0-12-821334-6.00003-X

effects of apathy and depression in individuals with probable AD. Participants conducted the depression criteria for people with dementia and the Neuropsychiatric Inventory (NPI). In a total of 734 participants with mild AD, 41.8% were found to have apathy with and without depression combined, and 47.8% of participants were found to have depression with and without apathy. Regarding depression specifically, 15.4% of the participants experienced depression, while only 9.4% of the participants experienced apathy alone, suggesting depression to be a common symptom of mild AD. The symptoms associated with depression were mainly fatigue and loss of energy, decreased positive affect or pleasure in social situations and activities, and agitation, among other symptoms (Benoit et al., 2012). The findings of Benoit et al. (2012) suggest that depression is a common experience of individuals with dementia that affects overall well-being, and negatively affects lifestyle if no intervention is implemented or available.

Further, several studies have shown that depression in dementia patients can lead to an increase in falls (Gostynski et al., 2001; Shua-Haim, Haim, Shi, Kuo, & Smith, 2001). Depression in AD can lead to greater isolation, higher morbidity as well as behavioral problems, such as physical and verbal aggression (Evers et al., 2002; Menon et al., 2001). Depression was also found to impact quality of life in older adults at risk of dementia, individuals with mild cognitive impairment (MCI), and patients with AD (Winter, Korchounov, Zhukova, & Bertschi, 2011), although some other studies did not find an effect of depression on quality of life (Nikmat, Hawthorne, & Al-Mashoor, 2011).

There are wide variations in reports of the prevalence of depression in AD and these variations have largely been attributed to inconsistent diagnostic criteria (Teng et al., 2008; Vilalta-Franch et al., 2006). Several studies argue that most assessments only use depression rating scales or subjective judgments and not standardized diagnostic criteria and clinical interviews (Byers & Yaffe, 2011; Wragg & Jeste, 1989). Studies of depression in AD vary not only in relation to estimations of prevalence, but also clinical correlates and response to treatment.

In one study, the prevalence, diagnosis, and treatment of depression was assessed using antemortem data for 279 dementia patients (type unspecified) and 24 nondemented healthy controls (Evers et al., 2002). Participants resided in chronic care facilities in their last 6 months of life. Documentation of mood disorders, depressive symptoms, pharmacologic management, and dementia severity was reviewed. This included clinical diagnosis of depression as determined by the International Classification of

Diseases-9 (ICD-9) codes for depressive disorders. It was found that major depression was prevalent in 29% of patients with dementia and 42% of nondemented controls. It was also found that depression was underdiagnosed among dementia patients, and recognition of depression was significantly lower for those with severe dementia relative to controls. Evers et al. (2002) also found that depression was undertreated in both groups. However, they were able to assess the impact of a depression screening program on the diagnosis and treatment of depression within the dementia population. The screening program involved nursing staff screening residents on admission and quarterly for mood symptoms of withdrawn or passive behavior, sad affect, lack of participation in activities, and sleep or appetite changes. Participants who reported these symptoms were referred to the facility's psychiatrist for evaluation. The documentation of depressive symptoms improved when comparing the cohort that had been included in the screening program with those who had not.

A review by Novais and Starkstein (2015) shows that there are significant discrepancies in estimates of frequency, main clinical correlates, and response to treatment of depression in AD. It was suggested that this may be partially explained by methodological limitations, lack of specific diagnostic criteria, and reliance on information provided by either caregivers or patients. A major difficulty in relation to the diagnosis of depression in AD was found to be the potential overlap between the symptoms of depression and symptoms resulting from cognitive decline in dementia. Such symptoms include insomnia, loss of interest, psychomotor retardation, and concentration deficits, as well as poor appetite, weight loss, and low self-esteem. Novais and Starkstein (2015) recommend that despite the fact that many criteria have not be properly validated, diagnosis of depression in AD should be based on systematic mental status examination leading to a psychiatric diagnosis.

Below we discuss several topics on the relationship between depression and dementia, including depression being a risk factor for the development of dementia, subtypes of depression in dementia, as well as how depressive symptoms manifest in different dementia groups.

Methods

In this study, we provide a narrative review on the relationship between depression and dementia. Our search strategy included a combination of two key words from two sets. The first set included depression, negative mood, or depressive symptoms. The second set included dementia, AD, vascular dementia, semantic dementia, and MCI. We have searched previous studies in PubMed, PsychoInfo, and also Google. Further, we have examined each paper carefully to make sure the goal of the study is studying depression in dementia. Studies that did not address this topic were excluded.

Depression as a risk factor for dementia

There are two potential directions of causal influence in depression and dementia: depression-to-dementia (i.e., depression can increase the prevalence of dementia) and dementia-to-depression [i.e., dementia symptoms can contribute to the development of depression (Thorpe (2009)]. Most existing studies focus on the depression-to-dementia pathway (Herbert & Lucassen, 2016; Jaroudi et al., 2017). For discussion on the latter point, see Barnes et al. (2012); Fischer et al. (2019); Moustafa et al. (2019); Moustafa, Tindle, Frydecka, & Misiak (2017); Russell-Williams et al. (2018); Zeki Al Hazzouri et al. (2014).

Depression-to-dementia is supported by evidence that depression is an early prodromal phase of dementia (for a review, see Muliyala & Varghese, 2010). Depression has also been identified as a potential risk factor for MCI or the progression of MCI to dementia due to AD (Lebedeva et al., 2017; Lee et al., 2012; Mah, Binns, Steffens, & Alzheimer's Disease Neuroimaging, 2015; Monastero, Mangialasche, Camarda, Ercolani, & Camarda, 2009). Indeed, depression has been shown to mediate the association between MCI and dementia (Devanand et al., 1996; Gabryelewicz et al., 2007; Li, Meyer, & Thornby, 2001a; 2001b; Simard, van Reekum, & Cohen, 2000; Teng, Lu, & Cummings, 2007; Vloeberghs, Opmeer, De Deyn, Engelborghs, & De Roeck, 2018). Several studies have shown that the occurrence of depression can at least double the likelihood of developing AD (Geerlings et al., 2000; Jorm, 2001; Modrego & Ferrandez, 2004; Ownby, Crocco, Acevedo, John, & Loewenstein, 2006; Palmer et al., 2007; Palmer et al., 2010). Even in patients with high educational levels, depression precedes memory decline later in life [Geerlings et al. (2000)].

It is important to note that the time when depression occurs over an individual's lifespan relates to the risk of developing dementia. The relationship between early- and late-life depression and the risk of dementia was explored by Li et al. (2011) using a longitudinal study. 3410 participants were involved in laboratory tests, neuroimaging tests, tests for dementia using the Diagnostic and Statistical Manual of Mental Disorders Version 4 (DSM-IV), and a measure of depression using the Center for Epidemiological Studies Depression (CES-D). Results show that depression can increase the risk of dementia, only when it occurred during late life. Depression occurring in individuals aged 50 years and over increases the risk of dementia, but depression occurring in individuals before the age of 50 years does not increase the risk of dementia. It is suggested that depression is a risk factor for the development of dementia (Li et al., 2011).

Another factor that influences the development of dementia is the frequency of depressive symptoms. Dotson, Beydoun, & Zonderman (2010) investigated whether the number of depressive episodes is related to the risk of developing dementia. Similar to Li et al. (2011), Dotson et al. (2010) conducted a longitudinal study over a period of roughly 24 years involving 1239 participants. Dementia progression and severity were measured throughout the study using the Clinical Dementia Rating (CDR), the Dementia Questionnaire, and the Diagnostic and Statistical Manual Version 3 Revised (DSM-III-R). Depression was also measured using the CES-D. Results show that 87% of the participants show increased risk of dementia when they had experienced one episode of depressive symptoms. Moreover, participants who experienced two depressive episodes had an even double risk of developing dementia. Furthermore, the findings highlighted that with each depressive episode, individuals had a 14% increased risk of developing dementia (Dotson et al., 2010). Such findings emphasize the necessity of behavioral and cognitive interventions to manage the symptoms and experiences of depression, which can have detrimental impacts on the development and experiences of dementia (Dotson et al., 2010; Harwood, Barker, Ownby, & Duara, 1999; Li et al., 2011; Zubenko et al., 2003).

Harwood et al. (1999) recruited 243 AD patients to explore whether there is a relationship between the history of depression and current depression in AD patients. Current depression was measured using the Cornell scale while patient and caregiver interviews also assessed the history of depression and mood disorder (e.g., suicide ideation, psychiatric hospitalization). It was revealed that a significantly more number of AD patients with a history of depression and mood disturbances had current depression relative to those with no history of depression. Therefore, it is argued that a history of depression increases the likelihood of mood disturbance, including depression, among AD patients (Zubenko, 1996).

To investigate the occurrence and clinical features of depression in AD patients, Zubenko et al. (2003) recruited probable AD patients and a comparison group of cognitively normal individuals. Several assessments of cognitive status and functioning were conducted by reviewing medical and behavioral disturbance history, as well as via physical examination, neuroimaging, and neuropsychological assessments [e.g., Mini-Mental State Examination (MMSE) and Clinical Dementia Rating Scale (CDRS)]. Major depression was assessed using the Clinical Assessment of Depression in Dementia (CADD), which is a diagnostic interview that incorporates segments of the Hamilton Depression Rating Scale (HDRS) and NPI scales related to the neuropsychiatric experiences of dementia patients such as hallucination and depressive symptoms. Overall, it was found that AD patients experiencing major depressive episodes tended to be younger when diagnosed with probable AD relative to patients without depression. In addition, the CADD revealed higher HDRS scores and more delusions and hallucinations among AD patients with major depression relative to AD patients who did not experience major depressive episodes. A possible interpretation of this finding is that major depression is an early sign of probable dementia. Moreover, AD patients had a higher rate of previous depressive episodes (44.9%) than the cognitively normal participants (28.5%). In the AD group, 18.5% had experienced depressive episodes before the diagnosis with the disease, suggesting that depression increases the risk of AD development, and the reoccurrence of depressive episodes after AD (34.6% occurrence rate).

The abovementioned studies show that depression can precede the development of dementia. These studies suggest that depression can impact hippocampal function, which subsequently leads to the development of dementia (Chung et al., 2016; Jaroudi et al., 2017; Steffens et al., 2002). However, some studies suggested that while depression may precede cognitive decline and memory problems in AD patients, it is not necessarily the case that depression is a risk factor for the development of AD (Bennett & Thomas, 2014; Chen, Ganguli, Mulsant, & DeKosky, 1999; Chen, Hu, Wei, Qin, & Copeland, 2009). That is, it is possible that depression can occur before the development of dementia without causing AD. Future animal studies may be able to provide some information on the causal relationship between depression and dementia. Further, as previously noted, there is also evidence for a dementia-to-depression pathway. Patients with dementia may develop depression due to their illness. Moreover, the prevalence of depression in dementia varies widely depending on the type of dementia, instruments used, and diagnostic definitions. Regardless of the underlying pathway, the comorbidity of dementia and depression causes an increase in deficits of functioning, problematic behavior, nursing-home placement, caregiver stress (Berger et al., 2005), and increased mortality.

Subtypes of depression in dementia

There are different kinds of depression: major versus minor depression as well as dysthymia. While major depression refers to formal definition of depression according to DSM-5, minor depression is a subclinical form of depression, which is defined as having 2–4 depressive symptoms, including either depressed mood or loss of interest (Fils et al., 2010). Dysthymia is defined as mild form of dementia but usually lasts longer (Niculescu & Akiskal, 2001).

The prevalence of minor and major depressive episodes (as defined by neuropsychiatric examination and using DSM-IV criteria) was assessed in 109 outpatients with AD (Lyketsos et al., 1997). All participants met the criteria for probable AD. The mean age of patients was 74.4 years [standard deviation (SD) 7.9]; 79% were women and 84% were living in their own home or that of a family member. 17% of participants had a history of depressive disorder prior to the onset of AD. All participants were rated on the Cornell Scale for Depression in Dementia (CSDD), the Psychogeriatric Dependency Rating Scale (subscales for physical dependency and behavioral disturbance), and the General Medical Health Rating. Participants were allocated into three groups based on their CSDD scores: major depressive episode (22% of participants), minor depressive episode (27% of participants), and "remaining patients". The presence of major depressive episode was associated with substantial impairments in activities of daily living (ADL), severe nonmood behavioral disturbance, and frequent wandering. Minor depressive episode was associated with moderately severe nonmood behavioral disturbance and frequent wandering. These findings support the findings of Teri and Wagner (1992) in that major depression in AD is associated with worse behavioral disturbance.

The prevalence, risk factors, and correlates of depression were examined in 103 patients who met clinical criteria for probable AD (Migliorelli et al., 1995). These patients were examined using a structured psychiatric interview assessing the presence of cognitive impairments, deficits in ADL, social functioning, and anosognosia. At least two first-degree relatives of the patients were interviewed with structured psychiatric evaluation. The following psychiatric tools were administered: Structured Clinical Interview for DSM-III-R (SCID), Structured Clinical Interview for DSM-III-R Personality Disorders (SCID-II), Present State Examination, Family History Research Diagnostic Criteria, HDRS, Hamilton Anxiety Rating Scale (HAM-A), Anosognosia Questionnaire-Dementia, Functional Independence Measure, and the Social Ties Checklist. The MMSE was also administered, along with an extensive list of neuropsychological examinations measuring visual perception, auditory attention, ideomotor apraxis, abstract reasoning, and other neurological mechanisms. It was shown that 51% of the patients had a depressive disorder (28% dysthymia and 23% major depression). The majority of patients with major depression or dysthymia were women, and those with major depression had significantly fewer years of formal education. Half of the patients with major depression had depression onset prior to the onset of dementia. This was in stark contrast to the patients with dysthymia, 86% of which had the onset of depression *after* the onset of dementia. Another significant finding was that patients with dysthymia had significantly less anosognosia (i.e., better awareness of their cognitive and behavioral problems) than in patients with major depressive disorder and healthy controls. Migliorelli et al. (1995) suggest that dysthymia could be related to having greater awareness of their progressive cognitive decline. Interestingly, only 6 patients (out of 53 suffering either dysthymia or major depression) were taking antidepressant medication. Patients with major depression had significantly higher HDRS scores than patients with dysthymia or no depression, while those with dysthymia had significantly higher HDRS scores than nondepressed patients. Migliorelli et al. (1995) also examined the prevalence of dysthymia and major depression in mild, moderate, and severe stages of dementia as defined by DSM-III-R criteria. They found that dysthymia was more prevalent among patients with mild dementia, while major depression was prevalent similarly across the progressive stages.

Chemerinski, Petracca, Sabe, Kremer, and Starkstein (2001) investigated the prevalence of "masked depression" in AD—a term used to describe a patient who meets DSM-IV criteria for major or minor depression in the absence of depressed mood or loss of interest (i.e., they meet four or more of the DSM criteria for either major or minor depression without the required initial symptom of depressed mood or loss of interest). There were 233 patients with AD, 47 with depression/without dementia, and 20 healthy controls. AD patients were then divided into two groups: nondepressed

Patients with AD and depressed patients with AD. The following instruments were used: the SCID, the MMSE, the HDRS, the HAM-A, the CDRS, and the Apathy Scale. Participants with AD who were given a score of 2+ in the "depressed mood" item of the HDRS by their caregiver were labeled "with depressed mood" ($n = 92$). Those who scored 0 on this item were labeled "without depressed mood" ($n = 62$). Depressed patients with AD had significantly higher scores on anxiety, apathy, and Parkinsonism than the nondepressed participants with AD. The depressed patients with AD had significantly higher score than the nondepressed patients with AD on the following HDRS items: guilt, suicide, middle insomnia, late insomnia, loss of interest, psychomotor retardation, agitation, worry, anxiety, loss of energy, loss of libido, hypochondriasis, and loss of weight. There was no significant difference regarding the loss of appetite. Relative to depressed patients with AD, those without AD had significantly higher scores on HDRS items: suicide, loss of appetite, and loss of weight than those with AD. However, those with AD had significantly higher scores on psychomotor retardation than those without AD. Further, 96% of depressed patients with AD (as defined by their score of 2+ on the "depressed mood" criteria of the HDRS by their caregiver) met DSM-IV criteria for either major (61%) or minor (35%) depression. Only 2% met the conditions for masked major depression, while 18% met the criteria for masked minor depression. The significant association between depressed mood and the large range of depressive symptoms suggests both affective and autonomic symptoms of depression are frequent among patients with AD. Further, these symptoms appear to be specific to a mood disorder, and not epiphenomenal to a chronic neurological disease.

Overall, several studies have shown that there are different types of depression in dementia patients: major depression, minor depression, and dysthymia. These types of depression have different effects on clinical aspects of dementia and memory decline. Further, the development of dementia was found to be more related to late-life rather than early- or midlife depression (Singh-Manoux et al., 2017). Thus it is suggested that clinicians consider the impact of different types of depression as well as whether it happened in early or late life on patients' well-being.

Depression and insight

Several studies have reported a relationship between insight and depression in dementia patients. Wilson, Schneider, Arnold, Bienias, and Bennett (2007) explored the relationship between personality traits, cognitive functioning, and individuals' risk of developing AD. Participants were Catholic priests, nuns, and religious brothers. Results show that symptoms of depression and awareness were related to the development and worsening of cognitive functions, such that the more aware individuals are of their disease (i.e., having higher insight), the more depressed they feel. This concept of depression occurring more in individuals with awareness of their disease is supported by literature. Brodaty and Luscombe (1996) found that depressive symptoms occur more frequently and intensely in individuals with dementia who have intact disease insight in comparison to individuals with impaired disease insight. In Brodaty and Luscombe (1996), participants with higher MMSE scores had lower scores on the HDRS, and vice versa. The results suggest that as cognitive impairment starts, depression reaches a peak and then starts to decrease as cognition and disease insight gets worse.

In support of the idea that there is a relationship between depression and disease insight, Harwood and Sultzer (2002) explored whether individuals with AD experience hopelessness, mood disturbances, and psychiatric and behavioral symptoms, which are key characteristics of depression. The researchers investigated the presence of these feelings concerning disease insight. A sample of 91 participants with AD completed the HDRS to measure for depression and hopelessness. They also completed the Neurobehavioral Rating Scale to measure psychiatric and behavioral mood disturbance and insight, and the MMSE to measure cognition. Results show that the majority of the participants had feelings of hopelessness and other negative thoughts as well as psychological symptoms of mood disturbance. These participants tended to have an awareness of their disease. Based on these findings, it can be said that depression affects individuals in their early stages of dementia as feelings of hopelessness can lead to suicide ideation (Harwood & Sultzer, 2002; Sabodash, Mendez, Fong, & Hsiao, 2013). These findings highlight that there is a relationship between disease insight and mood disturbance. More specifically, during early dementia when individuals are aware of their disease, and their worsening cognition and function, they experience more mood disturbance and develop depressive symptoms.

Depression in subtypes of dementia: relation to insight

The severity of depression varies across MCI, AD, and vascular dementia, as well as disease severity (e.g., mild vs severe dementia). While some assume that the severity of depression may increase as dementia symptoms become worse, this is not the case. According to Barca, Selbaek, Laks, and Engedal (2008), depression is more common in mild than severe dementia possibly due to greater illness insight. Along these lines, Barca et al. (2008) report that the prevalence of depressive symptoms is higher among patients with mild dementia (CDR 1) and severe dementia (CDR 3), compared to those without dementia or those with moderate dementia (CDR 2) due to psychological and biological risk factors (which are unspecified in the study).

One study investigated the prevalence of significant depressive symptoms in 270 individuals with MCI and 402 AD patients (Van der Mussele et al., 2013). Depressive symptoms were assessed using the CSDD. Behavioral assessment was completed by both the patients and their caregivers using the following scales over a 2-week period: The Middelheim Frontality Score (MFS), Behavioral Pathology in Alzheimer's Disease Rating Scale (Behave-AD), Cohen-Mansfield Agitation Inventory, and CSDD. The results show that depressive symptoms are higher in AD patients (25%) than in individuals with MCI (16%). AD patients with depressive symptoms displayed more severe frontal lobe symptoms, behavioral symptoms, and agitated behavior as compared with individuals with MCI with depressive symptoms. Severity of frontal lobe symptoms (using the MFS scale) was also higher in MCI individuals with depressive symptoms relative to MCI individuals without depression. Further, the severity of frontal lobe symptoms was higher in AD patients with depressive symptoms than AD patients without depression. Restlessness, impaired control of emotions, euphoria and emotional bluntness, disinhibition, aspontaneity, and stereotyped behavior were all more prevalent in AD patients with depression than without depression. According to the Behave-AD scale, 74% of AD patients with depressive symptoms presented with moderate to severe behavioral symptoms.

Lyketsos et al. (2000) recruited 1002 individuals: 329 with dementia (214 AD; 62 vascular dementia; 32 dementia with unknown etiology; 4 frontal temporal lobe dementia; 3 alcohol-related dementia; 2 posttraumatic stress disorder-related dementia). Further, the history of current mental, cognitive, and behavioral situation of participants was evaluated. Assessment included dementia severity and mental and behavioral evaluation using the CDR and NPI, considering experiences such as hallucination and depression. The results indicate that the behavioral disturbance of apathy, a domain of the NPI, is the most common in dementia patients followed by depression/aggression. Patients with vascular dementia had higher rates of depression than AD patients, while AD patients expressed higher rates of delusions than vascular dementia patients. Lyketsos et al. (2000) found depression and hallucinations decreased in severe dementia, which might be linked to the lack of awareness (or insight) in later stages of disease. However, it is possible that lack of awareness prevents patients with severe dementia from expressing their mood states. Thus future work should attempt to assess biological markers of depression in patients with mild versus severe dementia.

In another study, AD ($n = 111$) and semantic dementia ($n = 25$) patients were compared on suicidal behavior in relation to depression and insight (Sabodash et al., 2013). Information on the history and current state of participants' suicide ideation and depression was collected using a neuropsychiatric symptoms checklist, as well as some information on relevant family history. Cognitive functioning was evaluated using the MMSE and the mini-Boston Naming Test. Insight was measured with the University of California Los Angeles insight interview, which involved scaling patients' awareness of their disease/disability and how they expressed it (e.g., with or without concern or denial). It was found that most of the semantic dementia patients had a history of depression and suicidal behavior, and some of the patients had an occurrence of depression in the early stages of their disease. Moreover, it was found that semantic dementia patients have more depression and suicidal ideation than AD patients (two semantic dementia patients died by suicide). The authors argued that semantic dementia patients are more aware of their disease (i.e., have more insight) and think of it more as concerning, leading to increased depression and thoughts of hopelessness.

Depression occurs more in vascular dementia patients than in AD patients (Lyketsos et al., 2000), and the levels of insight and personality changes may also differ between these two types of dementia. Using CT scans, Verhey, Ponds, Rozendaal, and Jolles (1995) evaluated neuropathology in both AD and vascular dementia patients. They also assessed dementia severity using the GDS and the Brief Dementia Scale. The HDRS was used to assess depression, and a 4-point rating scale during an interview involving the patient and caregiver was used to identify discrepancies between patients' history and knowledge of their disease. They found that depression is more common in vascular dementia patients than in AD patients. This is consistent with Lyketsos et al. (2000), who suggest that vascular dementia manifests and affects brain regions associated with mood while AD pathology affects regions associated with cognition. However, Verhey et al. (1995) found that patients with mild stages of both diseases had similar levels of depression in comparison to patients in later stages of the diseases, although insight was not significantly different between the patient groups. This suggests that awareness of dementia (i.e., insight) may not be related to depression.

Unlike the abovementioned studies, one study has investigated the different subtypes of depression in different subtypes of dementia. Castilla-Puentes and Habeych (2010) investigated the prevalence of major depressive disorder, recurrent episode, dysthymic disorder, depressive psychosis, and adjustment disorder among 6440 patients with dementia (2947 with AD, 725 with vascular dementia, and 2768 with unspecified dementia). Subtypes of depression were identified using the Integrated Healthcare Information services database and subtypes of dementia were classified according to the ICD-9 diagnostic criteria. The prevalence of depressive disorder was 27% in all dementia patients. The vascular dementia group reported significantly higher prevalence rates in each of the depressive disorders when compared to both the AD and unspecified dementia group. Further, patients in the unspecified dementia group had significantly higher prevalence rates of depressive disorders when compared to the AD group. Patients between 60 and 64 years had a lower prevalence of depression compared with patients aged 78 years and older (26.2% vs 27.94%, respectively) and these differences persisted between the subgroups of dementia. The prevalence of depressive psychosis was low and comparable across the groups. Dysthymic disorder and adjustment disorder with depressive symptoms were significantly higher in the vascular dementia group compared with AD patients but not compared to unspecified dementia patients. No significant differences were found for dysthymic disorders between AD and unspecified dementia groups; however, significant differences were found in adjustment disorder between AD and unspecified dementia groups. The prevalence of dysthymia was found to be 2% in AD patients, 3% in unspecified dementia patients, and 4% in vascular dementia patients. In conclusion, this study suggests that depressive disorders are more prevalent in patients with vascular dementia than AD and unspecified dementia. Major depressive disorder, depressive disorder, and adjustment disorder with depressive features are more frequent in vascular dementia; however, the prevalence of dysthymia is similar between unspecified dementia and vascular dementia patients.

In sum, the abovementioned studies show that depression is more common in vascular and semantic dementia than in AD. Further, the prevalence of depression in dementia patients may be related to their level of insight (i.e., awareness of their disease). However, further research is therefore needed to clarify how insight might relate to the link between dementia and depression.

Conclusions and future work

Our review shows that the relationship between depression and dementia is complex. First, there have many conflicting results on the prevalence of depression in dementia. This has to do with the several scales used to measure depression, as well as relying on caregiver or clinician's assessment of the patients. Future work should attempt to integrate and provide a unified measure of depression in dementia. Furthermore, while several studies have found that depression precedes the development of dementia, it is not clear if depression does indeed cause dementia. This is because most of these studies are not controlled (which is difficult to do in human clinical studies). However, animal studies may be able to provide some information regarding whether depression can lead the development of dementia. Furthermore, we find that insight may be related to the experience of depression in dementia, such that dementia patients with intact insight are more likely to report depressive symptoms. Future work should provide biomarkers for depression in dementia that are not necessarily related to insight. Further, our review shows that there are several subtypes of depression in dementia and that the severity of depressive symptoms depends on dementia type, as depression was found to be more prevalent in patients with vascular and semantic dementia than in patients with AD. However, future work should explain the neural mechanism underlying the prevalence of depression in dementia, and why it is more common in vascular and semantic dementia than in AD.

References

Amore, M., Tagariello, P., Laterza, C., & Savoia, E. M. (2007). Subtypes of depression in dementia. *Archives of Gerontology and Geriatrics*, *44*(Suppl 1), 23–33. Available from https://doi.org/10.1016/j.archger.2007.01.004.

Barca, M. L., Selbaek, G., Laks, J., & Engedal, K. (2008). The pattern of depressive symptoms and factor analysis of the Cornell scale among patients in Norwegian nursing homes. *International Journal of Geriatric Psychiatry*, *23*(10), 1058–1065. Available from https://doi.org/10.1002/gps.2033.

Barnes, D. E., Yaffe, K., Byers, A. L., McCormick, M., Schaefer, C., & Whitmer, R. A. (2012). Midlife vs late-life depressive symptoms and risk of dementia: Differential effects for Alzheimer disease and vascular dementia. *Archives of General Psychiatry, 69* (5), 493–498. Available from https://doi.org/10.1001/archgenpsychiatry.2011.1481.

Baruch, N., Burgess, J., Pillai, M., & Allan, C. L. (2019). Treatment for depression comorbid with dementia. *Evidence-based Mental Health, 22*(4), 167–171. Available from https://doi.org/10.1136/ebmental-2019-300113.

Bennett, S., & Thomas, A. J. (2014). Depression and dementia: Cause, consequence or coincidence? *Maturitas, 79*(2), 184–190. Available from https://doi.org/10.1016/j.maturitas.2014.05.009.

Benoit, M., Berrut, G., Doussaint, J., Bakchine, S., Bonin-Guillaume, S., Fremont, P., & Robert, P. (2012). Apathy and depression in mild Alzheimer's disease: A cross-sectional study using diagnostic criteria. *Journal of Alzheimer's Disease: JAD, 31*(2), 325–334. Available from https://doi.org/10.3233/JAD-2012-112003.

Berger, G., Bernhardt, T., Weimer, E., Peters, J., Kratzsch, T., & Frolich, L. (2005). Longitudinal study on the relationship between symptomatology of dementia and levels of subjective burden and depression among family caregivers in memory clinic patients. *Journal of Geriatric Psychiatry and Neurology, 18*(3), 119–128. Available from https://doi.org/10.1177/0891988704273375.

Brodaty, H., & Luscombe, G. (1996). Depression in persons with dementia. *International Psychogeriatrics/IPA, 8*(4), 609–622.

Byers, A. L., & Yaffe, K. (2011). Depression and risk of developing dementia. *Nature Reviews Neurology,* 7(6), 323–331. Available from https://doi.org/10.1038/nrneurol.2011.60.

Castilla-Puentes, R. C., & Habeych, M. E. (2010). Subtypes of depression among patients with Alzheimer's disease and other dementias. *Alzheimer's & Dementia: The Journal of the Alzheimer's Association, 6*(1), 63–69. Available from https://doi.org/10.1016/j.jalz.2009.04.1232.

Chemerinski, E., Petracca, G., Sabe, L., Kremer, J., & Starkstein, S. E. (2001). The specificity of depressive symptoms in patients with Alzheimer's disease. *The American Journal of Psychiatry, 158*(1), 68–72. Available from https://doi.org/10.1176/appi.ajp.158.1.68.

Chen, P., Ganguli, M., Mulsant, B. H., & DeKosky, S. T. (1999). The temporal relationship between depressive symptoms and dementia: A community-based prospective study. *Archives of General Psychiatry, 56*(3), 261–266.

Chen, R., Hu, Z., Wei, L., Qin, X., & Copeland, J. R. (2009). Is the relationship between syndromes of depression and dementia temporal? The MRC-ALPHA and Hefei-China studies. *Psychological Medicine, 39*(3), 425–430. Available from https://doi.org/10.1017/S0033291708003735.

Chung, J. K., Plitman, E., Nakajima, S., Chakravarty, M. M., Caravaggio, F., Takeuchi, H., & Graff-Guerrero, A. (2016). Depressive symptoms and small hippocampal volume accelerate the progression to dementia from mild cognitive impairment. *Journal of Alzheimer's Disease: JAD, 49*(3), 743–754. Available from https://doi.org/10.3233/JAD-150679.

Devanand, D. P., Sano, M., Tang, M. X., Taylor, S., Gurland, B. J., Wilder, D., & Mayeux, R. (1996). Depressed mood and the incidence of Alzheimer's disease in the elderly living in the community. *Archives of General Psychiatry, 53*(2), 175–182.

Dotson, V. M., Beydoun, M. A., & Zonderman, A. B. (2010). Recurrent depressive symptoms and the incidence of dementia and mild cognitive impairment. *Neurology, 75*(1), 27–34. Available from https://doi.org/10.1212/WNL.0b013e3181e62124.

Evers, M. M., Samuels, S. C., Lantz, M., Khan, K., Brickman, A. M., & Marin, D. B. (2002). The prevalence, diagnosis and treatment of depression in dementia patients in chronic care facilities in the last six months of life. *International Journal of Geriatric Psychiatry*, *17*(5), 464–472. Available from https://doi.org/10.1002/gps.634.

Fils, J. M., Penick, E. C., Nickel, E. J., Othmer, E., Desouza, C., Gabrielli, W. F., & Hunter, E. E. (2010). Minor versus major depression: A comparative clinical study. *Primary Care Companion to the Journal of Clinical Psychiatry*, *12*(1). Available from https://doi.org/10.4088/PCC.08m00752blu, PCC 08m00752.

Fischer, A., Dourado, M. C. N., Laks, J., Landeira-Fernandez, J., Morris, R. G., & Mograbi, D. C. (2019). Modelling the impact of functionality, cognition, and mood state on awareness in people with Alzheimer's disease. *International Psychogeriatrics/IPA*, 1–11. Available from https://doi.org/10.1017/S1041610219001467.

Gabryelewicz, T., Styczynska, M., Luczywek, E., Barczak, A., Pfeffer, A., Androsiuk, W., & Barcikowska, M. (2007). The rate of conversion of mild cognitive impairment to dementia: Predictive role of depression. *International Journal of Geriatric Psychiatry*, *22*(6), 563–567. Available from https://doi.org/10.1002/gps.1716.

Geerlings, M. I., Schoevers, R. A., Beekman, A. T., Jonker, C., Deeg, D. J., Schmand, B., & Van Tilburg, W. (2000). Depression and risk of cognitive decline and Alzheimer's disease. Results of two prospective community-based studies in the Netherlands. *The British Journal of Psychiatry: The Journal of Mental Science*, *176*, 568–575.

Gostynski, M., Ajdacic-Gross, V., Heusser-Gretler, R., Gutzwiller, F., Michel, J. P., & Herrmann, F. (2001). [Dementia, depression and activity of daily living as risk factors for falls in elderly patients]. *Sozial- und Praventivmedizin*, *46*(2), 123–130.

Harwood, D. G., & Sultzer, D. L. (2002). "Life is not worth living": Hopelessness in Alzheimer's disease. *Journal of Geriatric Psychiatry and Neurology*, *15*(1), 38–43. Available from https://doi.org/10.1177/089198870201500108.

Harwood, D. G., Barker, W. W., Ownby, R. L., & Duara, R. (1999). Association between premorbid history of depression and current depression in Alzheimer's disease. *Journal of Geriatric Psychiatry and Neurology*, *12*(2), 72–75. Available from https://doi.org/10.1177/089198879901200206.

Herbert, J., & Lucassen, P. J. (2016). Depression as a risk factor for Alzheimer's disease: Genes, steroids, cytokines and neurogenesis—What do we need to know? *Frontiers in Neuroendocrinology*, *41*, 153–171. Available from https://doi.org/10.1016/j.yfrne.2015.12.001.

Jaroudi, W., Garami, J., Garrido, S., Hornberger, M., Keri, S., & Moustafa, A. A. (2017). Factors underlying cognitive decline in old age and Alzheimer's disease: The role of the hippocampus. *Reviews in the Neurosciences*, *28*(7), 705–714. Available from https://doi.org/10.1515/revneuro-2016-0086.

Jorm, A. F. (2001). History of depression as a risk factor for dementia: An updated review. *The Australian and New Zealand Journal of Psychiatry*, *35*(6), 776–781. Available from https://doi.org/10.1046/j.1440-1614.2001.00967.x.

Kuring, J. K., Mathias, J. L., & Ward, L. (2018). Prevalence of depression, anxiety and PTSD in people with dementia: A systematic review and meta-analysis. *Neuropsychology Review*, *28*(4), 393–416. Available from https://doi.org/10.1007/s11065-018-9396-2.

Lebedeva, A. K., Westman, E., Borza, T., Beyer, M. K., Engedal, K., Aarsland, D., & Haberg, A. K. (2017). MRI-based classification models in prediction of mild cognitive impairment and dementia in late-life depression. *Frontiers in Aging Neuroscience*, *9*, 13. Available from https://doi.org/10.3389/fnagi.2017.00013.

Lee, G. J., Lu, P. H., Hua, X., Lee, S., Wu, S., Nguyen, K., & Alzheimer's Disease Neuroimaging, I. (2012). Depressive symptoms in mild cognitive impairment predict greater atrophy in Alzheimer's disease-related regions. *Biological Psychiatry*, *71*(9), 814–821. Available from https://doi.org/10.1016/j.biopsych.2011.12.024.

Li, G., Wang, L. Y., Shofer, J. B., Thompson, M. L., Peskind, E. R., McCormick, W., & Larson, E. B. (2011). Temporal relationship between depression and dementia: Findings from a large community-based 15-year follow-up study. *Archives of General Psychiatry*, *68*(9), 970–977. Available from https://doi.org/10.1001/archgenpsychiatry.2011.86.

Li, Y. S., Meyer, J. S., & Thornby, J. (2001a). Longitudinal follow-up of depressive symptoms among normal versus cognitively impaired elderly. *International Journal of Geriatric Psychiatry*, *16*(7), 718–727.

Li, Y., Meyer, J. S., & Thornby, J. (2001b). Depressive symptoms among cognitively normal versus cognitively impaired elderly subjects. *International Journal of Geriatric Psychiatry*, *16*(5), 455–461.

Lyketsos, C. G., Steele, C., Baker, L., Galik, E., Kopunek, S., Steinberg, M., & Warren, A. (1997). Major and minor depression in Alzheimer's disease: Prevalence and impact. *The Journal of Neuropsychiatry and Clinical Neurosciences*, *9*(4), 556–561. Available from https://doi.org/10.1176/jnp.9.4.556.

Lyketsos, C. G., Steinberg, M., Tschanz, J. T., Norton, M. C., Steffens, D. C., & Breitner, J. C. (2000). Mental and behavioral disturbances in dementia: Findings from the cache county study on memory in aging. *The American Journal of Psychiatry*, *157*(5), 708–714. Available from https://doi.org/10.1176/appi.ajp.157.5.708.

Mah, L., Binns, M. A., Steffens, D. C., & Alzheimer's Disease Neuroimaging, I. (2015). Anxiety symptoms in amnestic mild cognitive impairment are associated with medial temporal atrophy and predict conversion to Alzheimer disease. *The American Journal of Geriatric Psychiatry: Official Journal of the American Association for Geriatric Psychiatry*, *23*(5), 466–476. Available from https://doi.org/10.1016/j.jagp.2014.10.005.

Menon, A. S., Gruber-Baldini, A. L., Hebel, J. R., Kaup, B., Loreck, D., Itkin Zimmerman, S., & Magaziner, J. (2001). Relationship between aggressive behaviors and depression among nursing home residents with dementia. *International Journal of Geriatric Psychiatry*, *16*(2), 139–146.

Migliorelli, R., Teson, A., Sabe, L., Petracchi, M., Leiguarda, R., & Starkstein, S. E. (1995). Prevalence and correlates of dysthymia and major depression among patients with Alzheimer's disease. *The American Journal of Psychiatry*, *152*(1), 37–44. Available from https://doi.org/10.1176/ajp.152.1.37.

Modrego, P. J., & Ferrandez, J. (2004). Depression in patients with mild cognitive impairment increases the risk of developing dementia of Alzheimer type: A prospective cohort study. *Archives of Neurology*, *61*(8), 1290–1293. Available from https://doi.org/10.1001/archneur.61.8.1290.

Monastero, R., Mangialasche, F., Camarda, C., Ercolani, S., & Camarda, R. (2009). A systematic review of neuropsychiatric symptoms in mild cognitive impairment. *Journal of Alzheimer's Disease: JAD*, *18*(1), 11–30. Available from https://doi.org/10.3233/JAD-2009-1120.

Moustafa, A. A., Crouse, J. J., Herzallah, M. M., Salama, M., Mohamed, W., Misiak, B., & Mattock, K. (2019). Depression following major life transitions in women: A review and theory. *Psychological Reports*. Available from https://doi.org/10.1177/0033294119872209, 33294119872209.

Moustafa, A. A., Tindle, R., Frydecka, D., & Misiak, B. (2017). Impulsivity and its relationship with anxiety, depression and stress. *Comprehensive Psychiatry*, *74*, 173–179. Available from https://doi.org/10.1016/j.comppsych.2017.01.013.

Muliyala, K. P., & Varghese, M. (2010). The complex relationship between depression and dementia. *Annals of Indian Academy of Neurology*, *13*(Suppl 2), S69–S73. Available from https://doi.org/10.4103/0972-2327.74248.

Niculescu, A. B., 3rd, & Akiskal, H. S. (2001). Proposed endophenotypes of dysthymia: Evolutionary, clinical and pharmacogenomic considerations. *Molecular Psychiatry*, *6*(4), 363–366. Available from https://doi.org/10.1038/sj.mp.4000906.

Nikmat, A. W., Hawthorne, G., & Al-Mashoor, S. H. (2011). Quality of life in dementia patients: Nursing home versus home care. *International Psychogeriatrics/IPA*, *23*(10), 1692–1700. Available from https://doi.org/10.1017/S1041610211001050.

Novais, F., & Starkstein, S. (2015). Phenomenology of depression in Alzheimer's disease. *Journal of Alzheimer's Disease: JAD*, *47*(4), 845–855. Available from https://doi.org/10.3233/JAD-148004.

Ownby, R. L., Crocco, E., Acevedo, A., John, V., & Loewenstein, D. (2006). Depression and risk for Alzheimer disease: Systematic review, meta-analysis, and metaregression analysis. *Archives of General Psychiatry*, *63*(5), 530–538. Available from https://doi.org/10.1001/archpsyc.63.5.530.

Palmer, K., Berger, A. K., Monastero, R., Winblad, B., Backman, L., & Fratiglioni, L. (2007). Predictors of progression from mild cognitive impairment to Alzheimer disease. *Neurology*, *68*(19), 1596–1602. Available from https://doi.org/10.1212/01.wnl.0000260968.92345.3f.

Palmer, K., Di Iulio, F., Varsi, A. E., Gianni, W., Sancesario, G., Caltagirone, C., & Spalletta, G. (2010). Neuropsychiatric predictors of progression from amnestic-mild cognitive impairment to Alzheimer's disease: The role of depression and apathy. *Journal of Alzheimer's Disease: JAD*, *20*(1), 175–183. Available from https://doi.org/10.3233/JAD-2010-1352.

Porter, V. R., Buxton, W. G., Fairbanks, L. A., Strickland, T., O'Connor, S. M., Rosenberg-Thompson, S., & Cummings, J. L. (2003). Frequency and characteristics of anxiety among patients with Alzheimer's disease and related dementias. *The Journal of Neuropsychiatry and Clinical Neurosciences*, *15*(2), 180–186. Available from https://doi.org/10.1176/jnp.15.2.180.

Rapp, M. A., Schnaider-Beeri, M., Grossman, H. T., Sano, M., Perl, D. P., Purohit, D. P., & Haroutunian, V. (2006). Increased hippocampal plaques and tangles in patients with Alzheimer disease with a lifetime history of major depression. *Archives of General Psychiatry*, *63*(2), 161–167. Available from https://doi.org/10.1001/archpsyc.63.2.161.

Reichman, W. E., & Coyne, A. C. (1995). Depressive symptoms in Alzheimer's disease and multi-infarct dementia. *Journal of Geriatric Psychiatry and Neurology*, *8*(2), 96–99. Available from https://doi.org/10.1177/089198879500800203.

Russell-Williams, J., Jaroudi, W., Perich, T., Hoscheidt, S., El Haj, M., & Moustafa, A. A. (2018). Mindfulness and meditation: Treating cognitive impairment and reducing stress in dementia. *Reviews in the Neurosciences*, *29*(7), 791–804. Available from https://doi.org/10.1515/revneuro-2017-0066.

Sabodash, V., Mendez, M. F., Fong, S., & Hsiao, J. J. (2013). Suicidal behavior in dementia: A special risk in semantic dementia. *American Journal of Alzheimer's Disease and Other Dementias*, *28*(6), 592–599. Available from https://doi.org/10.1177/1533317513494447.

Shua-Haim, J. R., Haim, T., Shi, Y., Kuo, Y. H., & Smith, J. M. (2001). Depression among Alzheimer's caregivers: Identifying risk factors. *American Journal of Alzheimer's Disease and Other Dementias*, *16*(6), 353–359. Available from https://doi.org/10.1177/153331750101600611.

Simard, M., van Reekum, R., & Cohen, T. (2000). A review of the cognitive and behavioral symptoms in dementia with Lewy bodies. *The Journal of Neuropsychiatry and Clinical Neurosciences*, *12*(4), 425–450. Available from https://doi.org/10.1176/jnp.12.4.425.

Singh-Manoux, A., Dugravot, A., Fournier, A., Abell, J., Ebmeier, K., Kivimaki, M., & Sabia, S. (2017). Trajectories of depressive symptoms before diagnosis of dementia: A 28-year follow-up study. *JAMA Psychiatry*, *74*(7), 712–718. Available from https://doi.org/10.1001/jamapsychiatry.2017.0660.

Australian Bureau of Statistics (2015). National Health Survey: First results (no. 4364.0.55.001). Retrieved from http://www.abs.gov.au/ausstats/abs@.nsf/Lookup/by%20Subject/4364.0.55.001~2014-15~Main%20Features~Mental%20and%20behavioural%20conditions~32.

Steffens, D. C., Payne, M. E., Greenberg, D. L., Byrum, C. E., Welsh-Bohmer, K. A., Wagner, H. R., & MacFall, J. R. (2002). Hippocampal volume and incident dementia in geriatric depression. *The American Journal of Geriatric Psychiatry: Official Journal of the American Association for Geriatric Psychiatry*, *10*(1), 62–71.

Teng, E., Lu, P. H., & Cummings, J. L. (2007). Neuropsychiatric symptoms are associated with progression from mild cognitive impairment to Alzheimer's disease. *Dementia and Geriatric Cognitive Disorders*, *24*(4), 253–259. Available from https://doi.org/10.1159/000107100.

Teng, E., Ringman, J. M., Ross, L. K., Mulnard, R. A., Dick, M. B., & Bartzokis, G. (2008). Diagnosing depression in Alzheimer disease with the national institute of mental health provisional criteria, Alzheimer's Disease Research Centers of California-Depression in Alzheimer's Disease, I.*The American Journal of Geriatric Psychiatry: Official Journal of the American Association for Geriatric Psychiatry*, *16*(6), 469–477. Available from https://doi.org/10.1097/JGP.0b013e318165dbae.

Teri, L., & Wagner, A. (1992). Alzheimer's disease and depression. *Journal of Consulting and Clinical Psychology*, *60*(3), 379–391.

Thorpe, L. (2009). Depression vs. dementia: How do we assess? *The Canadian Review of Alzheimer's Disease and Other Dementias*, *12*(3), 17–21.

Van der Mussele, S., Bekelaar, K., Le Bastard, N., Vermeiren, Y., Saerens, J., Somers, N., & Engelborghs, S. (2013). Prevalence and associated behavioral symptoms of depression in mild cognitive impairment and dementia due to Alzheimer's disease. *International Journal of Geriatric Psychiatry*, *28*(9), 947–958. Available from https://doi.org/10.1002/gps.3909.

Verhey, F. R., Ponds, R. W., Rozendaal, N., & Jolles, J. (1995). Depression, insight, and personality changes in Alzheimer's disease and vascular dementia. *Journal of Geriatric Psychiatry and Neurology*, *8*(1), 23–27.

Vilalta-Franch, J., Garre-Olmo, J., Lopez-Pousa, S., Turon-Estrada, A., Lozano-Gallego, M., Hernandez-Ferrandiz, M., & Feijoo-Lorza, R. (2006). Comparison of different clinical diagnostic criteria for depression in Alzheimer disease. *The American Journal of Geriatric Psychiatry: Official Journal of the American Association for Geriatric Psychiatry*, *14*(7), 589–597. Available from https://doi.org/10.1097/01.JGP.0000209396.15788.9d.

Vloeberghs, R., Opmeer, E. M., De Deyn, P. P., Engelborghs, S., & De Roeck, E. E. (2018). [Apathy, depression and cognitive functioning in patients with MCI and dementia]. *Tijdschrift Voor Gerontologie en Geriatrie*, *49*(3), 95–102. Available from https://doi.org/10.1007/s12439-018-0248-6.

Wilson, R. S., Schneider, J. A., Arnold, S. E., Bienias, J. L., & Bennett, D. A. (2007). Conscientiousness and the incidence of Alzheimer disease and mild cognitive impairment. *Archives of General Psychiatry*, *64*(10), 1204–1212. Available from https://doi.org/10.1001/archpsyc.64.10.1204.

Winter, Y., Korchounov, A., Zhukova, T. V., & Bertschi, N. E. (2011). Depression in elderly patients with Alzheimer dementia or vascular dementia and its influence on their quality of life. *Journal of Neurosciences in Rural Practice*, *2*(1), 27–32. Available from https://doi.org/10.4103/0976-3147.80087.

Wragg, R. E., & Jeste, D. V. (1989). Overview of depression and psychosis in Alzheimer's disease. *The American Journal of Psychiatry*, *146*(5), 577–587. Available from https://doi.org/10.1176/ajp.146.5.577.

Zeki Al Hazzouri, A., Vittinghoff, E., Byers, A., Covinsky, K., Blazer, D., Diem, S., & Yaffe, K. (2014). Long-term cumulative depressive symptom burden and risk of cognitive decline and dementia among very old women. *The Journals of Gerontology. Series A, Biological Sciences and Medical Sciences*, *69*(5), 595–601. Available from https://doi.org/10.1093/gerona/glt139.

Zubenko, G. S. (1996). Clinicopathologic and neurochemical correlates of major depression and psychosis in primary dementia. *International Psychogeriatrics/IPA*, *8*(Suppl 3), 219–223, discussion 269-272.

Zubenko, G. S., Zubenko, W. N., McPherson, S., Spoor, E., Marin, D. B., Farlow, M. R., & Sunderland, T. (2003). A collaborative study of the emergence and clinical features of the major depressive syndrome of Alzheimer's disease. *The American Journal of Psychiatry*, *160*(5), 857–866. Available from https://doi.org/10.1176/appi.ajp.160.5.857.

CHAPTER 3

Stress and anxiety in dementia

Ahmed A. Moustafa[1,2,3], Shimaa Adel Heikal[4], Wafa Jaroudi[5,6] and Ahmed Helal[7]

[1]School of Psychology, Western Sydney University, Sydney, NSW, Australia
[2]MARCS Institute for Brain and Behaviour, Western Sydney University, Sydney, NSW, Australia
[3]Department of Human Anatomy and Physiology, The Faculty of Health Sciences, University of Johannesburg, Johannesburg, South Africa
[4]Biotechnology Program, School of Science and Engineering, The American University in Cairo, New Cairo, Egypt
[5]MARCS Institute for Brain, Behaviour, and Development, Western Sydney University, Sydney, NSW, Australia
[6]School of Social Sciences and Psychology, Western Sydney University, Sydney, NSW, Australia
[7]Faculty of Education, Department of Mental Health, Tanta University, Egypt

Dementia background

The prevalence of dementia is increasing worldwide and it is anticipated to affect more than 130 million people by 2050 (Martin Prince et al., 2015). In Australia, the number of individuals diagnosed with dementia has been also increasing (Health Direct, 2018). Dementia is a syndrome that includes various diseases that affect one's memory and cognition, and interfere with occupational and social processes [Gale, Acar, & Daffner, 2018; World Health Organization (WHO), 2017]. The WHO categorized dementia as a public health priority in 2012 because it significantly impoverished individuals' daily functioning as a consequence of a deterioration in memory, cognition, and behavior. It is predominantly chronic and progressive, and is induced by multiple brain illnesses (Baumgart et al., 2015; Ferri et al., 2005; WHO, 2012). The number of people with dementia is predicted to continue growing, especially in countries with demographic transition, as well as in older populations (WHO, 2012). Aging is impacting the prevalence of dementia and driving governments to develop action plan responses as the disease is a leading contributor to disability and dependence in older people (Prince et al., 2013).

In 2010, approximately 35.6 million individuals had dementia, suggesting 7.7 million individuals are being diagnosed with dementia each year worldwide (WHO, 2012). The total number of people diagnosed with dementia is projected to triple to more than 152 million by 2050 (WHO, 2018). The financial burden of dementia care was also estimated at $818 billion in 2015 with lower costs per person in the lower-income countries; the cost per person was estimated as $32865 in high-income countries, $6827 in upper-middle income

Alzheimer's Disease
DOI: https://doi.org/10.1016/B978-0-12-821334-6.00006-5

countries, $3109 in lower-middle income countries, and $868 in low-income countries (WHO, 2015, 2017). Although there is an increased number of young-onset cases of dementia, the disease is usually associated with old age (Prince et al., 2013; Prince, 2000). The older an individual gets, the higher their chances are of developing dementia [Australian Bureau of Statistics (ABS), 2015a].

Dementia is the second leading cause of death in individuals over 85 years in Australia (Australian Bureau of Statistics, 2015a; Waite, Broe, Grayson, & Creasey, 2001). The ABS reports that in 2015 there were 1,59,052 dementia-related deaths. Studies indicate that in 9 years, 5,89,000 individuals will be diagnosed with dementia, and this number will increase to over 1 million in 39 years in Australia (Health Direct, 2018). These numbers might increase as studies indicate that approximately 40% of patients with dementia are undiagnosed due to the intrinsic difficulties in differentiating the natural decline in overall cognitive functioning as a result of age from abnormal neurocognitive impairment (Slavin, Brodaty, & Sachdev, 2013). The statistics show a staggering increase in dementia over time and highlight the importance of research on the topic. Understanding the symptoms of dementia is crucial in order to efficiently develop interventions that alleviate the consequences of the disease and increase the patients' quality of life.

The decline in cognitive abilities is generally the main reported symptom in older patients. However, there is a typical degree of cognitive slowing associated with aging (Hugo & Ganguli, 2014). When the cognitive decline symptoms progress to a degree that hinders everyday functionality, dementia diagnosis would be a typical case. The intermediate stage between normal cognition and dementia is known as mild cognitive impairment (MCI) (Hugo & Ganguli, 2014). Impaired memory is one of the most predominant clues of developing cognitive impairment, which later progresses into dementia. More specifically, Holzer and Warshaw (2000) suggest a decline in memory recall is one of the earliest signs of dementia. Other symptoms include engaging in repetitive behaviors and showing worsened judgment, which can have a significant impact in their day-to-day activities such as driving or maintaining conversations (Aretouli & Brandt, 2010; Holzer & Warshaw, 2000).

Mental health in dementia

Mental health problems include anxiety and depression (Mental Health Foundation, 2016). The Australian Bureau of Statistics indicates that between 2014 and 2015, 2.1 million individuals had mental health issues associated with depression and anxiety. Moreover, psychological distress, which is the prolonged feeling of stress, is also reported as prevalent among the Australian population (Australian Bureau of Statistics, 2015c). Based on the current literature, there is a suggested link between dementia and mental health, in particular, depression and anxiety, and sometimes stress. It is also necessary to explore how each mood state negatively impacts the lives of individuals with dementia, particularly, daily living and quality of life to enlighten possible thoughts and actions that could improve the current rise of mental health and dementia.

Many of the mental health problems associated with dementia can become debilitating and interfere with the quality of life of individuals, and their ability to complete tasks required in daily living. Since Alzheimer's disease (AD) is the most common type of dementia that accounts for approximately 60%–80% of all dementia cases (Alzheimer's association, 2020), understanding its associated risk factors (e.g., depression and anxiety) would also help delaying the onset of dementia (Santabárbara et al., 2020). For example, anxiety was indicated to affect the progression of dementia (Becker et al., 2018). Accordingly, studying the role of mental health problems in the incidence of AD is of interest (Terracciano et al., 2013; Zufferey et al., 2017).

Many individuals diagnosed with dementia develop depression, anxiety, and stress as comorbid disorders (Bierman, Comijs, Jonker, & Beekman, 2007; Dementia Australia, 2015; Piccininni, Di Carlo, Baldereschi, Zaccara, & Inzitari, 2005). Depression, anxiety, and stress can contribute to cognitive impairment and dementia development, and are common experiences in individuals with dementia, especially during the mild and moderate stages of the disease (Zubenko, 1996). These three negative affect moods are generally overlooked and assumed to be normal experiences in old age and dementia. A reason as to why depression is overlooked or misdiagnosed as aging or dementia is due, in part, to the considerable overlap between mood disturbance symptoms and dementia.

Importantly, the limited cognitive capacity of individuals with dementia makes it even more challenging to identify and measure these negative mood states (Alzheimer's Association, 2018; Brodaty & Luscombe, 1996; Forstl et al., 1992; Kim & Rovner, 1994; Teresi, Abrams, Holmes, Ramirez, & Eimicke, 2001). Individuals with dementia have poor memory, concentration, and language comprehension, which make it challenging to complete measures of depression, anxiety, and stress. As a result, mental health issues tend to be underdiagnosed in care homes, especially by nurses and nurse aides (Teresi et al., 2001). The misdiagnosis and underdiagnosis of mood disturbances are important to address because negative moods, such as depression, can increase the risk and occurrence of suicide (Harwood & Sultzer, 2002; Sabodash, Mendex, Fong, & Hsiao, 2013). The Depression Anxiety Stress Scale (DASS-21) (Lovibond & Lovibond, 1995) is, to our knowledge, the only all-in-one psychological scale that measures depression, anxiety, and stress, but it cannot be used to test negative affect in individuals with cognitive impairment. The DASS-21 excludes individuals with cognitive impairment, such as those with dementia.

In addition to stress, anxiety, and depression, Van Der Linde, Dening, Matthews, and Brayne (2014) identified apathy to be a main reoccurring psychological symptom of dementia, which overlaps with depression and dysphoria (Van Der Linde et al., 2014). Both involve dissatisfaction or unhappiness with life. Apathy was reported to be highly prevalent and persistent throughout the course of dementia (van der Linde et al., 2016); the frequency of reported apathy was highly prevalent in cases with frontotemporal dementia (FTD), primary progressive aphasia (PPA), and young-onset AD (Collins, Henley, & Suárez-González, 2020).

In addition to apathy, Wilson, Krueger, et al. (2007) identified loneliness as a contributor to the development and experience of depression, stress, and AD. In addition, Wilson, Krueger, et al. (2007) conducted a longitudinal study involving cognitively healthy elderly participants. They found that loneliness was a contributing factor to depression, stress, and AD, and that it increased by twofold the risk of late-onset AD.

Wilson, Begeny, Boyle, Schneider, and Bennett (2011) explored what aspects of neuroticism—a personality trait that is known for the tendency to feel negative emotions, and involves experiences of anxiety, worry, fear, anger, frustration, guilt, and depressed mood—are associated with developing dementia. In a total of 785 elderly individuals without dementia, the two aspects of neuroticism generally associated with developing this condition were anxiety and stress. Concerning aspects of neuroticism and the development of dementia, depression was found to negatively impact specific cognitive functions of information processing and information retainment, such as memory (Wilson et al., 2011). The study findings suggest that depression, anxiety, and stress affect the development and progression of cognitive processes and diseases such as dementia can worsen cognitive functions such as memory (Wilson et al., 2011).

Another recent study investigated whether personality traits associate with differences in individual cognitive state. Zufferey et al. (2017) quantified the interaction between personality traits and cognitive impairment within the medial temporal lobe using hierarchical multivariate linear models. A significant interaction was found between personality traits and cognitive state especially neuroticism and the associated anxiety, depression, and stress.

Anxiety and dementia

Individuals with dementia commonly develop anxiety. Anxiety is a common symptom of dementia and distress and is associated with deteriorated quality of life as well as worsened neuropsychological performance (Hoe, Hancock, Livingston, & Orrell, 2006). It is estimated that dementia patients who have been diagnosed with anxiety are more than 25% (Seignourel, Kunik, Snow, Wilson, & Stanley, 2008). Anxiety disorders are also suggested to have a two- to four-time chance of occurring in individuals with dementia than in healthy elderly individuals (Chemerinski, Petracca, Manes, Leiguarda, & Starkstein, 1998). In addition, the development of anxiety in people with dementia leads to nursing home placement, which accordingly increases the caregivers' burden (Gibbons et al., 2002; Seignourel et al., 2008). Therefore addressing anxiety symptoms in patients with dementia has come into focus to suggest future approaches of disease management. In the next section, we discuss studies related to anxiety as a risk factor for dementia, its prevalence within different types of dementia, and anxiety in early versus late stages of dementia.

Anxiety as a risk factor for dementia

Several studies have investigated how anxiety impacts memory and dementia symptoms in healthy individuals and patients with dementia. For example, Wilson et al. (2005) investigated the relationship between distress and cognitive function in elderly individuals. Distress, which involves symptoms of anxiety and sorrow, is experienced differently by males and females. Female participants experience more distress than males, and this contributes to the former having a greater decline in cognitive performance (Wilson et al., 2005). This finding is in line with the notion that experiences of anxiety and sadness worsen overall cognitive abilities.

To elaborate on the relationship between distress, anxiety, and sadness, Wilson, Schneider, Arnold, Bienias, and Bennett (2007) and Wilson, Schneider, Boyle, et al. (2007) investigated the relationship between prolonged experiences of distress and the development of MCI. After 12 years of data collection, it was found that continual distress affects episodic memory, and increases the chance of developing MCI over time. Distress and anxiety were found to contribute to early cognitive decline. Therefore it is necessary to appropriately monitor and manage symptoms early on, as well as promote the implementation of efficacious interventions in elderly populations.

In one study on the association of anxiety with dementia in older men (48–67 years old), it was found that anxiety is a risk factor for the development of dementia (Gallacher et al., 2009). More recent systematic review and meta-analysis focused on anxiety as a predictor of dementia and cognitive decline. The results suggested an increased risk of cognitive impairment associated with anxiety, particularly in older adults (Gulpers et al., 2016).

Subjective cognitive decline (SCD) refers to one's own perception of cognitive ability. Liew (2020) measured anxiety symptoms in participants with SCD annually to explore any occurrence of dementia or MCI. SCD and anxiety proved to be risk factors for MCI and dementia with the highest probability associated with the cooccurrence of both symptoms (Liew, 2020). Moreover, Norman et al. (2020) studied the fear of developing Alzheimer's disease (FDAD) to understand how it can affect older peoples' subjective memory complaints (SMC). The study included adults aged 65–93 years, who self-reported symptoms of depression, anxiety, and memory, and the researchers rated their fear of developing AD. The conclusion was that anxiety is related to FDAD and SMC. Assessing anxiety symptoms in older people may help identify those who are at risk of developing apathy and distress (Norman et al., 2020).

Accordingly, early intervention is important as it identifies the experiences of anxiety and focuses on alleviating the symptoms associated with it, which, in turn, can reduce the likelihood of anxiety and cognitive impairment, thereby improving the quality of life of individuals with dementia.

The prevalence of anxiety in different types of dementia

Different studies have more closely investigated the occurrence of anxiety in different types of dementia. For example, Porter et al. (2003) evaluated the prevalence of anxiety in different dementia types, such as AD, vascular dementia (VaD), FTD, and compared it to the prevalence of anxiety experienced by healthy control participants. Data from 191 participants with dementia and 40 control participants were collected using the Neuropsychiatric Inventory (NPI) to measure anxiety and behavioral disturbance, the Mini-Mental State Examination (MMSE) to measure cognitive decline, the Functional Activities Questionnaire to measure ability to do daily living activities, and Magnetic Resonance Imaging brain scans and blood tests for hormone measures. Overall, participants with dementia were found to have more anxiety than control participants. When comparing dementia types, participants with VaD and FTD had higher anxiety than participants with AD. It is also interesting to note that participants with AD who experienced anxiety had early-onset dementia, which is dementia that occurs before the age of 65 (Porter et al., 2003). Such findings suggest that anxiety can increase the risk of developing dementia, and make the experiences of living with dementia more mentally fatiguing.

In line with Ballard et al. (2000), Porter et al. (2003) found that individuals with VaD experience anxiety symptoms more than individuals with AD. Ballard et al. (2000) reported that this is because of the differentiating disease pathologies of VaD and AD, such as their underlying development and their effects. Moreover, the results of Porter et al. (2003) imply a correlation between anxiety and cognition. That is, anxiety affects cognition and occurs most frequently in individuals who develop early-onset dementia (Porter et al., 2003). Therefore there could be a pattern between anxiety and dementia in which anxiety is a risk factor for developing dementia and is at peak during the mild stages of the disease.

Similar to Porter et al. (2003), Teri et al. (1999) explored the occurrence of anxiety symptoms and their relationship with possible comorbidities with depression. To do that, 523 participants with dementia had their cognitive abilities measured using the MMSE and the Mattis Dementia Rating Scale (Teri et al., 1999). Participants' ability to complete daily activities was also measured using the Blessed Dementia Rating Scale. Results show that 70% of the participants had one or more symptoms of anxiety, fear being the most predominant symptom. Other common behavioral symptoms that participants experienced included wandering and hallucinations (Teri et al., 1999).

Recently, a systematic review reported that anxiety is a risk factor for both AD and VaD. Rasmussen et al. (2018) reported anxiety and depression as independent risk factors in patients with AD and FTD. They reported a significant association between anxiety and FTD while AD was more associated with depression (Rasmussen et al., 2018). In addition, a retrospective study with AD and dementia with Lewy Bodies (DLB) sample of patients found anxiety was more frequent in DLB than in AD, after controlling for age, gender, and cognitive status. The symptoms appeared 4–5 years prior to clinical diagnosis, and was characterized by severe panic attacks that required medical help. Patients with AD were less likely to suffer from anxiety or at least never required medical help. Accordingly, the study suggested that DLB should be considered in all old patients admitted with anxiety and cognitive decline (Segers, Benoit, Meyts, & Surquin, 2020).

On the other hand, a sample of patients with AD and DLB studied over a 4-year period showed no significant anxiety associated with cognitive decline or disease severity. This suggests that anxiety is not an important factor in patients initially diagnosed with dementia (Breitve et al., 2016). In one study, neuropsychiatric symptoms associated with different types of dementia varied among the studies. Depression was highly prevalent in AD and FTD while anxiety was frequently reported in FTD (Collins et al., 2020).

Anxiety in early versus late stages of dementia

Several studies have measured anxiety in relation to different stages of dementia. For example, Seignourel et al. (2008) suggest that anxiety most commonly affects individuals in the early dementia stages due to their insight of their disease. Research has compared patterns of anxiety and depression in individuals at different stages of cognitive decline and AD, and found that in early stages of decline, patients exhibited many anxiety and depression symptoms, yet as the condition progressed toward an AD diagnosis, these mood symptoms decreased (Bierman et al., 2007). An initial awareness of the decline provoked increased anxiety and depressive symptoms in patients, while as the condition progressed and the awareness decreased, the feelings of anxiousness and apathy were less pronounced. Based on these findings, individuals who are in most need of mental health assessment and assistance are individuals with cognitive impairment and who have mild to moderate dementia. Individuals in the early stages of dementia are mostly affected by the disease in terms of mental health as they progress from being cognitively healthy to impaired.

Similarly, Piccininni et al. (2005) investigated the frequency of behavioral and psychological symptoms in AD to better understand the severity of such symptoms across the progression of the disease. A sample of 50 individuals with AD and their caregivers participated in the study, which included an evaluation of their mental health, neuropsychological health, and neuroimaging. The tests included scales to measure language functioning, memory processes, and the ability to do daily tasks. To measure participants' behavioral and psychological symptom severity of delusion, anxiety, irritability, and motor activity, caregivers were interviewed using the NPI. Results included increased apathy, aberrant motor activity, such as pacing and compulsions, dysphoria and anxiety to be most reported in AD participants, occurring 46% – 74% of the time, which is very frequent (Piccininni et al., 2005). Moreover, Piccininni et al. (2005) found that symptoms of delusions, hallucinations, and aberrant motor activity became more frequent and severe as the disease progressed. On the other hand, psychological symptoms, such as dysphoria were most frequent and severe in the mild stages of the disease when individuals were aware of their declining abilities, also known as having intact disease insight (Piccininni et al., 2005). Dysphoria is accompanied by feelings of depression, anxiety, and agitation, and commonly occurs in mild dementia, which highlights the importance of being able to evaluate and measure negative affect in individuals with dementia.

Stress and dementia

In addition to anxiety, stress is a mood state linked to cognitive impairment in individuals with dementia (Greenberg, Tanev, Marin, & Pitman, 2014; Mohlenhoff, O'Donovan, Weiner, & Neylan, 2017). Stress is known to be a factor that increases the risk of developing cognitive impairment. Johansson et al. (2010) explored the link between the experiences of stress and the development of dementia. Over a period of 35 years, participants' stress was monitored by regularly completing psychological stress measures administered by a physician. Neuropsychiatric examinations and the Diagnostic and Statistical Manual of Mental Disorders were also used to screen and evaluate individuals for dementia. It was found that ongoing midlife stress was associated with an increased risk of developing dementia, and experiences of stress were associated with both early- and late-onset dementia (Johansson et al., 2010).

Physiological responses to stress can affect the brain, and this may increase the risk of dementia (Islamoska et al., 2020; Ouanes & Popp, 2019). The effect of stress on the brain and its link to dementia is not yet fully understood; however, several mechanisms have been suggested including that elevated cortisol levels due to prolonged stress can lead to a dysregulation of the hypothalamus–pituitary–adrenal axis resulting in increased stress reactivity (Ouanes & Popp, 2019). In this section, we discuss how psychological stress affects memory and cognition, psychosocial stress in dementia, posttraumatic stress disorder (PTSD) in dementia, perceived midlife stress and dementia, childhood stress and dementia, and stress as experienced by dementia patients.

Psychological stress, MCI, and dementia

Psychological stress is an emerging risk factor for the development of MCI and dementia. Many studies have investigated the relation between experiencing psychological stress and developing cognitive decline in later life (Comijs, van den Kommer, Minnaar, Penninx, & Deeg, 2011; Johansson et al., 2010; Tschanz et al., 2013; Wilson, Krueger, et al., 2007). Nevertheless, the exact relationship between stress and cognitive outcomes is still unclear.

Katz et al. (2016) investigated stress and its relation to cognition, specifically the development of amnestic mild cognitive impairment (aMCI) using a longitudinal study. The study found stress to be a determinant of the development of aMCI. An important finding was that female participants with aMCI and significant levels of stress also showed high ratings of depression. Based on these findings, individuals with high-stress scores on the Perceived Stress Scale measure were calculated to have a two and a half-time increased chance of developing aMCI.

Psychosocial problems such as job strain, high demands, and low job control have been shown to cause individual stress (Karasek, 1979; Wang, Wahlberg, Karp, Winblad, & Fratiglioni, 2012). Research shows that psychosocial stress in the workplace can be a significant contributor to developing dementia (Seidler et al., 2004). Factors such as low job control and job strain are considered to be risk factors in the development of dementia and AD independent of other known factors (Wang et al., 2012).

Childhood stress, midlife stress, and dementia

Various studies have investigated childhood experiences and linked them to later adulthood health outcomes and disorders. Suffering abuse or being raised in a socioeconomic disadvantaged environment is associated with later-life psychiatric disorders (McCrory, Dooley, Layte, & Kenny, 2015). Experiencing stressful events during early life such as the loss of a parent increases the risk of dementia (Conde-Sala & Garre-Olmo, 2020; Norton, Østbye, Smith, Munger, & Tschanz, 2009; Radford et al., 2017). Moreover, childhood stress can include living in an orphanage, in custody, or experiencing crisis or war. The more stress undergone during childhood, the higher the risk of suffering dementia, which reinforces the role childhood experiences play in later-life consequences (Donley, Lönnroos, Tuomainen, & Kauhanen, 2018). Therefore special support should be implemented to children suffering from stress in order to protect them from possibly developing dementia or AD in adulthood.

As for midlife stress, a 35-yearlong longitudinal study found that females who developed dementia frequently reported suffering high and continuous stress in their middle age (Johansson et al., 2010). A 27-year follow-up study indicated that the incidence of dementia increased by 50% in participants who had previously reported mental distress (Skogen, Bergh, Stewart, Knudsen, & Bjerkeset, 2015). Age was found to be a significant factor because participants who reported distress during their midlife period were more likely to develop dementia compared to those who experienced stress in later adulthood.

Posttraumatic stress disorder in dementia

PTSD is a common form of stress that refers to a severe clinical disorder that arises after an individual experiences a life-threatening event, such as war or abuse (Flatt, Gilsanz, Quesenberry, Albers, & Whitmer, 2018; Mawanda, Wallace, McCoy, & Abrams, 2017). PTSD was identified as a potential risk factor for the development of dementia (Ball et al., 2009; Bruneau, Desmarais, & Pokrzywko, 2020; Günak et al., 2020; Kuring, Mathias, & Ward, 2020; Lawrence et al., 2020; Nilaweera et al., 2020). Wang et al. (2016) suggest that individuals with PTSD have a four-time increased risk of dementia development. Such findings highlight the negative impact that prolonged feelings and experiences of stress have on cognition, especially if the type of stress involves traumatic experiences, such as war or abuse (Uddo, Vasterling, Brailey, & Sutker, 1993; Weniger, Lange, Sachsse, & Irle, 2008). Katz et al. (2016) suggest that stress is an experience of individuals with dementia and can also increase the chances of developing MCI. Such findings emphasize the need for an intervention that can alleviate stress, and in turn, reduce the risk of developing MCI. Yaffe et al. (2010) support the idea of stress leading to the development of cognitive problems, which can later develop into dementia. Yaffe et al. (2010) investigated the development of dementia in a sample of veterans with and without PTSD. The findings showed that veterans with PTSD show higher chances of developing dementia than veteran participants without PTSD. Their findings also reveal that stress and PTSD lead to a shrinkage of the hippocampus, an area important for memory processes. Moreover, stress and PTSD lead to worsened or impaired memory and learning, such as short-term memory. Individuals who experience stress also have high cortisol levels, a chemical released in response to stress. These individuals also develop neural atrophy in the hippocampus (Duman, 2002). Therefore it is possible that the consequences of stress are more specific than increasing the risk of developing dementia as it explicitly impacts the memory functions within the hippocampus.

Flatt, Quesenberry, Liu, Albers, and Whitmer (2016) examined PTSD in men and women and its association with dementia. They reported that dementia was more prevalent in participants with PTSD than those without PTSD. In addition, regardless of gender, individuals who suffered PTSD were at risk of developing dementia (J. Flatt *et al.* 2016). Flatt et al. (2016) also showed that persons with PTSD and comorbid depression are at twofold risk of dementia (Flatt et al., 2018). Not only early-life PTSD was investigated, but also delayed-onset PTSD was associated with dementia. Moreover, Ball et al. (2009) investigated the increased aggression in patients with dementia and PTSD. Patients with coexisting dementia and PTSD were shown to be more aggressive but more studies are needed to support this hypothesis (Ball et al., 2009).

The notion of stress contributing to cognitive impairment is further confirmed by the finding that individuals with PTSD have smaller hippocampal volumes than normal healthy individuals. Brain scans also show that individuals with PTSD have hippocampal atrophy, which is associated with cell degeneration, leading to impaired memory recall (Uddo et al., 1993; Weniger et al., 2008). To expand on the idea of stress relating to memory impairment, cortisol levels also negatively affect synaptic plasticity and the form and structure of dendrites in the hippocampus (Kim & Diamond, 2002; McEwen, 2000).

These results indicate that the earlier recognition of a person's trauma would be beneficial as they will receive the suitable psychological interventions that help protect them from dementia and possible cognitive decline (Martinez-Clavera, James, Bowditch, & Kuruvilla, 2017).

Conclusions

Our review shows that anxiety is a risk factor for the development of dementia as it impacts memory and cognition. It appears that anxiety is more common in VaD than in AD. Similarly, like anxiety, stress also increases the chances of developing MCI and dementia. Our review also showed that both childhood and midlife stress increase the chances of developing dementia. Lastly, we have found a link between PTSD and the development of dementia.

Interventions that work on both cognitive and mental health aspects affected by dementia may have a two-way stream of assistance for individuals with this condition. That is, interventions that improve cognitive function can also improve mental health, and similarly, interventions that improve mental health can also improve cognition. It is therefore essential that, during the development of interventions, researchers bear in mind cognitive and mental health factors, including individuals' personality, such as neuroticism, as well as their previous experience with anxiety or stress and existing comorbidities such as depression. The proper identification of anxiety or stress symptoms in early-stage dementia patients would be necessary for personalized treatment and effective psychological care as a means of slowing down the progression of this illness. Finally, public health interventions should aim at reducing chronic stress, implement suitable care for people who have suffered childhood stress and PTSD, as well as provide a better overall social environment for persons at risk of dementia.

References

Alzheimer's Association. (2018). *Depression*. Retrieved from https://www.alz.org/help-

Alzheimer's association. (2020). 2020 Alzheimer's disease facts and figures. *Alzheimer's & Dementia, 16*(3), 391–460. Available from https://doi.org/10.1002/alz.12068.

Aretouli, E., & Brandt, J. (2010). Everyday functioning in mild cognitive impairment and its relationship with executive cognition. *International Journal of Geriatric Psychiatry: A Journal of the Psychiatry of Late Life and Allied Sciences, 25*(3), 224–233. Available from https://doi.org/10.1002/gps.2325.

Australian Bureau of Statistics. (2015a). *Causes of Death, Australia* (no. 3303.0). Retrieved from http://www.abs.gov.au/ausstats/abs@.nsf/Lookup/by%20Subject/3303.0~2015~Main%20Features~Dementia~10002.

Australian Bureau of Statistics. (2015b). *National Health Survey: First Results* (no.4364.0.55.001). Retrieved from http://www.abs.gov.au/ausstats/abs@.nsf/Lookup/by%20Subject/4364.0.55.001~2014-15~Main%20Features~Mental%20and%20behavioural%20conditions~32

Australian Bureau of Statistics. (2015c). *National Health Survey: First Results* (no. 4364.0.55.001). Retrieved from http://www.abs.gov.au/ausstats/abs@.nsf/Lookup/by%20Subject/4364.0.55.001~2014-15~Main%20Features~Psychological%20distress~16.

Ball, V. L., Hudson, S., Davila, J., Morgan, R., Walder, A., Graham, D. P., ... Kunik, M. E. (2009). Post-traumatic stress disorder and prediction of aggression in persons with dementia. *International Journal of Geriatric Psychiatry*, *24*(11), 1285–1290. Available from https://doi.org/10.1002/gps.2258.

Ballard, C., Neill, D., O'brien, J., McKeith, I. G., Ince, P., & Perry, R. (2000). Anxiety, depression and psychosis in vascular dementia: Prevalence and associations. *Journal of Affective Disorders*, *59*(2), 97–106. Available from https://doi.org/10.1016/S0165-0327(99)00057-9.

Baumgart, M., Snyder, H. M., Carrillo, M. C., Fazio, S., Kim, H., & Johns, H. (2015). Summary of the evidence on modifiable risk factors for cognitive decline and dementia: A population-based perspective. *Alzheimer's & Dementia*, *11*(6), 718–726. Available from https://doi.org/10.1016/j.jalz.2015.05.016.

Becker, E., Orellana Rios, C. L., Lahmann, C., Rücker, G., Bauer, J., & Boeker, M. (2018). Anxiety as a risk factor of Alzheimer's disease and vascular dementia. *British Journal of Psychiatry*, *213*(5), 654–660. Available from https://doi.org/10.1192/bjp.2018.173.

Bierman, E. J. M., Comijs, H. C., Jonker, C., & Beekman, A. T. F. (2007). Symptoms of anxiety and depression in the course of cognitive decline. *Dementia and Geriatric Cognitive Disorders*, *24*(3), 213–219. Available from https://doi.org/10.1159/000107083.

Breitve, M. H., Hynninen, M. J., Brønnick, K., Chwiszczuk, L. J., Auestad, B. H., Aarsland, D., & Rongve, A. (2016). A longitudinal study of anxiety and cognitive decline in dementia with Lewy bodies and Alzheimer's disease. *Alzheimer's Research and Therapy*, *8*(1), 3. Available from https://doi.org/10.1186/s13195-016-0171-4.

Benoit, M., Berrut, G., Doussaint, J., Bakchine, S., Bonin-Guillaume, S., Frémont, P., ... Sellal, F. (2012). Apathy and depression in mild Alzheimer's disease: A cross-sectional study using diagnostic criteria. *Journal of Alzheimer's Disease*, *31*(2), 325–334. Available from https://doi.org/10.3233/JAD-2012-112003.

Brodaty, H., & Luscombe, G. (1996). Depression in persons with dementia. *International Psychogeriatrics*, *8*(4), 609–622. Available from https://doi.org/10.1017/S104161029600292X.

Bruneau, M., Desmarais, P., & Pokrzywko, K. (2020). Post-traumatic stress disorder mistaken for behavioural and psychological symptoms of dementia: Case series and recommendations of care. *Psychogeriatrics*, *20*(5), 754–759. Available from https://doi.org/10.1111/psyg.12549.

Chemerinski, E., Petracca, G., Manes, F., Leiguarda, R., & Starkstein, S. E. (1998). Prevalence and correlates of anxiety in Alzheimer's disease. *Depression and Anxiety*, 7 (4), 166–170, 10.1002/(SICI)1520-6394(1998)7:4 < 166::AID-DA4 > 3.0.CO;2-8.

Collins, J. D., Henley, S. M. D., & Suárez-González, A. (2020). *A systematic review of the prevalence of depression, anxiety, and apathy in frontotemporal dementia, atypical and young-onset Alzheimer's disease, and inherited dementia. International Psychogeriatrics*. Cambridge University Press. Available from https://doi.org/10.1017/S1041610220001118.

Comijs, H. C., van den Kommer, T. N., Minnaar, R. W. M., Penninx, B. W. J. H., & Deeg, D. J. H. (2011). Accumulated and differential effects of life events on cognitive decline in older persons: Depending on depression, baseline cognition, or ApoE epsilon4 status? *The Journals of Gerontology. Series B, Psychological Sciences and Social Sciences*, *66*(Suppl 1). Available from https://doi.org/10.1093/geronb/gbr019.

Conde-Sala, J. L., & Garre-Olmo, J. (2020). Early parental death and psychosocial risk factors for dementia: A case–control study in Europe. *International Journal of Geriatric Psychiatry*, *35*(9), 1051–1059. Available from https://doi.org/10.1002/gps.5328.

Dementia Australia. (2015). Don't forget dementia on World Mental Health Day. Retrieved from https://www.dementia.org.au/media-releases/2015/dont-forget-dementia-on-world-mental-health-day.

Donley, G. A. R., Lönnroos, E., Tuomainen, T. P., & Kauhanen, J. (2018). Association of childhood stress with late-life dementia and Alzheimer's disease: The KIHD study. *European Journal of Public Health, 28*(6), 1069–1073. Available from https://doi.org/10.1093/eurpub/cky134.

Duman, R. S. (2002). Pathophysiology of depression: the concept of synaptic plasticity. *European Psychiatry, 17*, 306–310. Available from https://doi.org/10.1016/S0924-9338(02)00654-5.

Ferri, C. P., Prince, M., Brayne, C., Brodaty, H., Fratiglioni, L., Ganguli, M., ... Scazufca, M. (2005). Global prevalence of dementia: A Delphi consensus study. *Lancet, 366*(9503), 2112–2117. Available from https://doi.org/10.1016/S0140-6736(05)67889-0.

Flatt, J., Quesenberry, C. P., Liu, J. Y., Albers, K., & Whitmer, R. A. (2016). 01-02-03: Post-traumatic stress disorder and risk of dementia among men and women members of a healthcare delivery system. *Alzheimer's & Dementia, 12*. Available from https://doi.org/10.1016/j.jalz.2016.06.300, P174–P174.

Flatt, J. D., Gilsanz, P., Quesenberry, C. P., Albers, K. B., & Whitmer, R. A. (2018). Post-traumatic stress disorder and risk of dementia among members of a health care delivery system. *Alzheimer's & Dementia, 14*(1), 28–34. Available from https://doi.org/10.1016/j.jalz.2017.04.014.

Forstl, H., Burns, A., Luthert, P., Cairns, N., Lantos, P., & Levy, R. (1992). Clinical and neuropathological correlates of depression in Alzheimer's disease. *Psychological Medicine, 22*(4), 877–884. Available from https://doi.org/10.1017/S0033291700038459.

Gale, S. A., Acar, D., & Daffner, K. R. (2018). Dementia. *American Journal of Medicine, 131*(10), 1161–1169. Available from https://doi.org/10.1016/j.amjmed.2018.01.022, Elsevier Inc.

Gallacher, J., Bayer, A., Fish, M., Pickering, J., Pedro, S., Dunstan, F., ... Ben-Shlomo, Y. (2009). Does anxiety affect risk of dementia? Findings from the Caerphilly Prospective Study. *Psychosomatic Medicine, 71*(6), 659–666. Available from https://doi.org/10.1097/PSY.0b013e3181a6177c.

Gibbons, L., Teri, L., Logsdon, R., McCurry, S., Kukull, W., Bowen, J., ... Larson, E. (2002). Anxiety symptoms as predictors of nursing home placement in patients with Alzheimer's disease. *Journal of Clinical Geropsychology, 8*(4), 335–342. Available from https://doi.org/10.1023/A:1019635525375.

Gloster, A. T., Rhoades, H. M., Novy, D., Klotsche, J., Senior, A., Kunik, M., ... Stanley, M. A. (2008). Psychometric properties of the Depression Anxiety and Stress Scale-21 in older primary care patients. *Journal of Affective Disorders, 110*(3), 248–259. Available from https://doi.org/10.1016/j.jad.2008.01.023.

Greenberg, M. S., Tanev, K., Marin, M.-F., & Pitman, R. K. (2014). Stress, PTSD, and dementia. *Alzheimer's & Dementia, 10*(3 Suppl.), S155–S165. Available from https://doi.org/10.1016/j.jalz.2014.04.008.

Gulpers, B., Ramakers, I., Hamel, R., Köhler, S., Oude Voshaar, R., & Verhey, F. (2016). Anxiety as a predictor for cognitive decline and dementia: A systematic review and meta-analysis. *American Journal of Geriatric Psychiatry, 24*(10), 823–842. Available from https://doi.org/10.1016/j.jagp.2016.05.015, Elsevier B.V.

Günak, M. M., Billings, J., Carratu, E., Marchant, N. L., Favarato, G., & Orgeta, V. (2020). Post-traumatic stress disorder as a risk factor for dementia: Systematic review and meta-analysis. *The British Journal of Psychiatry: The Journal of Mental Science, 217*(5), 600–608. Available from https://doi.org/10.1192/bjp.2020.150.

Harwood, D. G., & Sultzer, D. L. (2002). "Life is not worth living": Hopelessness in Alzheimer's disease. *Journal of Geriatric Psychiatry and Neurology, 15*(1), 38–43. Available from https://doi.org/10.1177/089198870201500108.

Health Direct. (2018). *Dementia Statistics*. Retrieved from https://www.healthdirect.gov.au/dementia-statistics.

Hoe, J., Hancock, G., Livingston, G., & Orrell, M. (2006). Quality of life of people with dementia in residential care homes. *British Journal of Psychiatry, 188*(May), 460–464. Available from https://doi.org/10.1192/bjp.bp.104.007658.

Holzer, C., & Warshaw, G. (2000). Clues to early Alzheimer dementia in the outpatient setting. *Archives of Family Medicine, 9*(10), 1066, Retrieved from. Available from https://triggered.clockss.org/ServeContent?url = http://archfami.ama-assn.org%2Fcgi%2Freprint%2F9%2F10%2F1066.pdf.

Hori, K., Oda, T., Asaoka, T., Yoshida, M., Watanabe, S., Oyamada, R., ... Inada, T. (2005). First episodes of behavioral symptoms in Alzheimer's disease patients at age 90 and over, and early-onset Alzheimer's disease: Comparison with senile dementia of Alzheimer's type. *Psychiatry and Clinical Neurosciences, 59*(6), 730–735. Available from https://doi.org/10.1111/j.1440-1819.2005.01444.x.

Hugo, J., & Ganguli, M. (2014). Dementia and cognitive impairment. Epidemiology, diagnosis, and treatment. *Clinics in Geriatric Medicine, 30*(3), 421–442. Available from https://doi.org/10.1016/j.cger.2014.04.001, W.B. Saunders.

Islamoska, S., Ishtiak-Ahmed, K., Hansen, Å. M., Grynderup, M. B., Mortensen, E. L., Garde, A. H., ... Nabe-Nielsen, K. (2019). Vital exhaustion and incidence of dementia: Results from the Copenhagen City Heart Study. *Journal of Alzheimer's Disease, 67*(1), 369–379. Available from https://doi.org/10.3233/JAD-180478.

Islamoska, S., Hansen, Å. M., Ishtiak-Ahmed, K., Garde, A. H., Andersen, P.K., Garde, E., ... Nabe-Nielsen, K. (2020). Stress diagnoses in midlife and risk of dementia: A register-based follow-up study. Aging & Mental Health, 1–10. Available from https://doi.org/10.1080/13607863.2020.1742656.

Johansson, L., Guo, X., Waern, M., Östling, S. Ö., Gustafson, D., Bengtsson, C., & Skoog, I. (2010). Midlife psychological stress and risk of dementia: A 35-year longitudinal population study. *Brain, 133*(8), 2217–2224. Available from https://doi.org/10.1093/brain/awq116.

Karasek, R. A. (1979). Job demands, job decision latitude, and mental strain: Implications for job redesign. *Administrative Science Quarterly, 24*(2), 285. Available from https://doi.org/10.2307/2392498.

Katz, M. J., Derby, C. A., Wang, C., Sliwinski, M. J., Ezzati, A., Zimmerman, M. E., ... Lipton, R. B. (2016). Influence of perceived stress on incident amnestic mild cognitive impairment: Results from the Einstein Aging Study. *Alzheimer Disease and Associated Disorders, 30*(2), 93. Available from https://doi.org/10.1097/WAD.0000000000000125.

Kim, E., & Rovner, B. W. (1994). Depression in dementia. *Psychiatric Annals, 24*(4), 173–177. Available from https://doi.org/10.3928/0048-5713-19940401-06.

Kim, J. J., & Diamond, D. M. (2002). The stressed hippocampus, synaptic plasticity and lost memories. *Nature Reviews. Neuroscience, 3*(6), 453. Available from https://doi.org/10.1038/nrn849.

Kuring, J. K., Mathias, J. L., & Ward, L. (2020). Risk of dementia in persons who have previously experienced clinically-significant depression, anxiety, or PTSD: A systematic review and meta-analysis. *Journal of Affective Disorders, 274*, 247–261. Available from https://doi.org/10.1016/j.jad.2020.05.020, Elsevier B.V.

Lawrence, K. A., Pachner, T. M., Long, M. M., Henderson, S., Schuman, D. L., & Plassman, B. L. (2020). Risk and protective factors of dementia among adults with post-traumatic stress disorder: A systematic review protocol. *British Medical Journal Open, 10*(6). Available from https://doi.org/10.1136/bmjopen-2019-035517, e035517.

Li, G., Wang, L. Y., Shofer, J. B., Thompson, M. L., Peskind, E. R., McCormick, W., ... Larson, E. B. (2011). Temporal relationship between depression and dementia: Findings from a large community-based 15-year follow-up study. *Archives of General Psychiatry*, *68*(9), 970–977. Available from https://doi.org/10.1001/archgenpsychiatry.2011.86.

Liew, T. M. (2020). Subjective cognitive decline, anxiety symptoms, and the risk of mild cognitive impairment and dementia. *Alzheimer's Research and Therapy*, *12*(1), 107. Available from https://doi.org/10.1186/s13195-020-00673-8.

Lovibond, S. H., & Lovibond, P. F. (1995). *Manual for the Depression Anxiety & Stress Scales* (2nd ed.). Sydney: Psychology Foundation.

Martin Prince, A., Wimo, A., Guerchet, M., Gemma-Claire Ali, M., Wu, Y.-T., Prina, M., ... Xia, Z. (2015). *World Alzheimer Report 2015, The Global Impact of Dementia: An analysis of prevalence, incidence, cost and trends*. Retrieved from www.alz.co.uk/worldreport2015corrections.

Martinez-Clavera, C., James, S., Bowditch, E., & Kuruvilla, T. (2017). Delayed-onset post-traumatic stress disorder symptoms in dementia. *Progress in Neurology and Psychiatry*, *21*(3), 26–31. Available from https://doi.org/10.1002/pnp.477.

Mawanda, F., Wallace, R. B., McCoy, K., & Abrams, T. E. (2017). PTSD, psychotropic medication use, and the risk of dementia among US veterans: A Retrospective cohort study. *Journal of the American Geriatrics Society*, *65*(5), 1043–1050. Available from https://doi.org/10.1111/jgs.14756.

McCrory, C., Dooley, C., Layte, R., & Kenny, R. A. (2015). The lasting legacy of childhood adversity for disease risk in later life. *Health Psychology*, *34*(7), 687–696. Available from https://doi.org/10.1037/hea0000147.

McEwen, B. S. (2000). The neurobiology of stress: From serendipity to clinical relevance. *Brain Research*, *886*(1-2), 172–189. Available from https://doi.org/10.1016/S0006-8993(00)02950-4.

Mental Health Foundation. (2016). *The interface between dementia and mental health: An evidence review policy paper 2016*.

Mohlenhoff, B. S., O'Donovan, A., Weiner, M. W., & Neylan, T. C. (2017). Dementia risk in posttraumatic stress disorder: The relevance of sleep-related abnormalities in brain structure, amyloid, and inflammation. *Current Psychiatry Reports*, *19*(11), 89, Current Medicine Group LLC 1. Available from https://doi.org/10.1007/s11920-017-0835-1.

Nabe-Nielsen, K., Rod, N. H., Hansen, Å. M., Prescott, E., Grynderup, M. B., Islamoska, S., ... Westendorp, R. G. J. (2020). Perceived stress and dementia: Results from the Copenhagen city heart study. *Aging & Mental Health*, *24*(11), 1828–1836. Available from https://doi.org/10.1080/13607863.2019.1625304.

Nilaweera, D., Freak-Poli, R., Ritchie, K., Chaudieu, I., Ancelin, M. L., & Ryan, J. (2020). The long-term consequences of trauma and posttraumatic stress disorder symptoms on later life cognitive function and dementia risk. *Psychiatry Research*, *294*, 113506. Available from https://doi.org/10.1016/j.psychres.2020.113506.

Norman, A. L., Woodard, J. L., Calamari, J. E., Gross, E. Z., Pontarelli, N., Socha, J., ... Armstrong, K. (2020). The fear of Alzheimer's disease: Mediating effects of anxiety on subjective memory complaints. *Aging and Mental Health*, *24*(2), 308–314. Available from https://doi.org/10.1080/13607863.2018.1534081.

Norton, M. C., Østbye, T., Smith, K. R., Munger, R. G., & Tschanz, J. T. (2009). Early parental death and late-life dementia risk: Findings from the Cache County Study. *Age and Ageing*. Available from https://doi.org/10.1093/ageing/afp023.

Ouanes, S., & Popp, J. (2019). High cortisol and the risk of dementia and Alzheimer's disease: A review of the literature. *Frontiers in Aging Neuroscience*, *11*. Available from https://doi.org/10.3389/fnagi.2019.00043.

Piccininni, M., Di Carlo, A., Baldereschi, M., Zaccara, G., & Inzitari, D. (2005). Behavioral and psychological symptoms in Alzheimer's disease: Frequency and relationship with duration and severity of the disease. *Dementia and Geriatric Cognitive Disorders*, *19*(5-6), 276–281. Available from https://doi.org/10.1159/000084552.

Porter, V. R., Buxton, W. G., Fairbanks, L. A., Strickland, T., O'Connor, S. M., Rosenberg-Thompson, S., & Cummings, J. L. (2003). Frequency and characteristics of anxiety among patients with Alzheimer's disease and related dementias. *The Journal of Neuropsychiatry and Clinical Neurosciences*, *15*(2), 180–186. Available from https://doi.org/10.1176/jnp.15.2.180.

Prince, M. (2000). Methodological issues for population-based research into dementia in developing countries: A position paper from the 10/66 Dementia Research Group. *International Journal of Geriatric Psychiatry*, *15*(1), 21–30, https://doi.org/10.1002/(SICI)1099-1166(200001)15:1 < 21::AID-GPS71 > 3.0.CO;2-5.

Prince, M., Bryce, R., Albanese, E., Wimo, A., Ribeiro, W., & Ferri, C. P. (2013). The global prevalence of dementia: A systematic review and metaanalysis. *Alzheimer's & Dementia*, *9*(1), 63–75. Available from https://doi.org/10.1016/j.jalz.2012.11.007, Elsevier Inc.

Potkins, D., Myint, P., Bannister, C., Tadros, G., Chithramohan, R., Swann, A., ... Margallo-Lana, M. (2003). Language impairment in dementia: impact on symptoms and care needs in residential homes. *International Journal of Geriatric Psychiatry*, *18*(11), 1002–1006. Available from https://doi.org/10.1002/gps.1002.

Radford, K., Delbaere, K., Draper, B., Mack, H. A., Daylight, G., Cumming, R., ... Broe, G. A. (2017). Childhood stress and adversity is associated with late-life dementia in aboriginal Australians. *American Journal of Geriatric Psychiatry*, *25*(10), 1097–1106. Available from https://doi.org/10.1016/j.jagp.2017.05.008.

Raglio, A., Bellelli, G., Traficante, D., Gianotti, M., Ubezio, M. C., Gentile, S., ... Trabucchi, M. (2010). Efficacy of music therapy treatment based on cycles of sessions: A randomised controlled trial. *Aging and Mental Health*, *14*(8), 900–904. Available from https://doi.org/10.1080/13607861003713158.

Rasmussen, H., Rosness, T. A., Bosnes, O., Salvesen, Ø., Knutli, M., & Stordal, E. (2018). Anxiety and depression as risk factors in frontotemporal dementia and Alzheimer's disease: The HUNT study. *Dementia and Geriatric Cognitive Disorders Extra*, *8*(3), 414–425. Available from https://doi.org/10.1159/000493973.

Sabodash, V., Mendex, M. F., Fong, S., & Hsiao, J. J. (2013). Suicidal behavior in dementia: A special risk in semantic dementia. *American Journal of Alzheimer's Disease & Other Dementias*, *28*(6), 592–599. Available from https://doi.org/10.1177/1533317513494447.

Santabárbara, J., Lipnicki, D. M., Bueno-Notivol, J., Olaya-Guzmán, B., Villagrasa, B., & López-Antón, R. (2020). Updating the evidence for an association between anxiety and risk of Alzheimer's disease: A meta-analysis of prospective cohort studies. *Journal of Affective Disorders*. Available from https://doi.org/10.1016/j.jad.2019.11.065, Elsevier B.V.

Segers, K., Benoit, F., Meyts, J., & Surquin, M. (2020). Anxiety symptoms are quantitatively and qualitatively different in dementia with Lewy bodies than in Alzheimer's disease in the years preceding clinical diagnosis. *Psychogeriatrics*, *20*(3), 242–246. Available from https://doi.org/10.1111/psyg.12490.

Seidler, A., Nienhaus, A., Bernhardt, T., Kauppinen, T., Elo, A. L., & Frölich, L. (2004). Psychosocial work factors and dementia. *Occupational and Environmental Medicine*, *61*(12), 962–971. Available from https://doi.org/10.1136/oem.2003.012153.

Seignourel, P. J., Kunik, M. E., Snow, L., Wilson, N., & Stanley, M. (2008). Anxiety in dementia: A critical review. *Clinical Psychology Review*, *28*(7), 1071–1082. Available from https://doi.org/10.1016/j.cpr.2008.02.008.

Skogen, J. C., Bergh, S., Stewart, R., Knudsen, A. K., & Bjerkeset, O. (2015). Midlife mental distress and risk for dementia up to 27 years later: The Nord-Trøndelag Health Study (HUNT) in linkage with a dementia registry in Norway. *BMC Geriatrics*, *15*(1). Available from https://doi.org/10.1186/s12877-015-0020-5.

Slavin, M. J., Brodaty, H., & Sachdev, P. S. (2013). Challenges of diagnosing dementia in the oldest old population. *Journal of Gerontology, Series A: Biological Sciences and Medical Science*, *68*(9), 1103–1111. doi:10.1093/gerona/glt051.

Sussams, R., Schlotz, W., Clough, Z., Amin, J., Simpson, S., Abbott, A., … Holmes, C. (2020). Psychological stress, cognitive decline and the development of dementia in amnestic mild cognitive impairment. *Scientific Reports*, *10*(1). Available from https://doi.org/10.1038/s41598-020-60607-0.

Teresi, J., Abrams, R., Holmes, D., Ramirez, M., & Eimicke, J. (2001). Prevalence of depression and depression recognition in nursing homes. *Social Psychiatry and Psychiatric Epidemiology*, *36*(12), 613–620. Available from https://doi.org/10.1007/s127-001-8202-7.

Teri, L., Ferretti, L. E., Gibbons, L. E., Logsdon, R. G., McCurry, S. M., Kukull, W. A., … Larson, E. B. (1999). Anxiety in Alzheimer's disease: Prevalence and comorbidity. *Journals of Gerontology Series A: Biomedical Sciences and Medical Sciences*, *54*(7), M348–M352. Available from https://doi.org/10.1093/gerona/54.7.M348.

Terracciano, A., Iacono, D., O'Brien, R. J., Troncoso, J. C., An, Y., Sutin, A. R., … Resnick, S. M. (2013). Personality and resilience to Alzheimer's disease neuropathology: A prospective autopsy study. *Neurobiology of Aging*, *34*(4), 1045–1050. Available from https://doi.org/10.1016/j.neurobiolaging.2012.08.008.

Tschanz, J. T., Pfister, R., Wanzek, J., Corcoran, C., Smith, K., Tschanz, B. T., … Norton, M. C. (2013). Stressful life events and cognitive decline in late life: Moderation by education and age. The Cache County Study. *International Journal of Geriatric Psychiatry*, *28*(8), 821–830. Available from https://doi.org/10.1002/gps.3888.

Uddo, M., Vasterling, J. J., Brailey, K., & Sutker, P. B. (1993). Memory and attention in combat-related post-traumatic stress disorder (PTSD). *Journal of Psychopathology and Behavioral Assessment*, *15*(1), 43–52. Available from https://doi.org/10.1007/BF00964322.

Van Der Linde, R. M., Dening, T., Matthews, F. E., & Brayne, C. (2014). Grouping of behavioural and psychological symptoms of dementia. *International Journal of Geriatric Psychiatry*, *29*(6), 562–568. Available from https://doi.org/10.1002/gps.4037.

van der Linde, R. M., Dening, T., Stephan, B. C. M., Prina, A. M., Evans, E., & Brayne, C. (2016). Longitudinal course of behavioural and psychological symptoms of dementia: Systematic review. *British Journal of Psychiatry*, *209*(5), 366–377. Available from https://doi.org/10.1192/bjp.bp.114.148403.

Waite, L. M., Broe, A., Grayson, D. A., & Creasey, H. (2001). The incidence of dementia in an Australian community population: The Sydney older persons study. *International Journal of Geriatric Psychiatry*, *16*(7), 680–689. Available from https://doi.org/10.1002/gps.404.

Wang, H.-X., Wahlberg, M., Karp, A., Winblad, B., & Fratiglioni, L. (2012). Psychosocial stress at work is associated with increased dementia risk in late life. *Alzheimer's & Dementia*, *8*(2), 114–120. Available from https://doi.org/10.1016/j.jalz.2011.03.001.

Wang, T. Y., Wei, H. T., Liou, Y. J., Su, T. P., Bai, Y. M., Tsai, S. J., … Chen, M. H. (2016). Risk for developing dementia among patients with posttraumatic stress disorder: A nationwide longitudinal study. *Journal of Affective Disorders*, *205*, 306–310. Available from https://doi.org/10.1016/j.jad.2016.08.013.

Weniger, G., Lange, C., Sachsse, U., & Irle, E. (2008). Amygdala and hippocampal volumes and cognition in adult survivors of childhood abuse with dissociative disorders. *Acta Psychiatrica Scandinavica, 118*(4), 281–290. Available from https://doi.org/10.1111/j.1600-0447.2008.01246.x.

WHO. (2012). *Dementia: a public health priority*. Retrieved from https://apps.who.int/iris/bitstream/handle/10665/75263/9789241564458_eng.pdf;jsessionid = 91F60603FF10908D2230F876DE20819C?sequence = 1.

WHO. (2015). *WHO|Thematic briefs for the First WHO Ministerial Conference on Global Action Against Dementia, March 16–17, 2015*. WHO. Retrieved from https://www.who.int/mental_health/neurology/dementia/thematic_briefs_dementia/en/.

WHO. (2017). *WHO|Global action plan on the public health response to dementia 2017 – 2025*. WHO.

WHO. (2018). *WHO|Towards a dementia plan: A WHO guide*. WHO.

Wilson, R. S., Begeny, C. T., Boyle, P. A., Schneider, J. A., & Bennett, D. A. (2011). Vulnerability to stress, anxiety, and development of dementia in old age. *The American Journal of Geriatric Psychiatry, 19*(4), 327–334. Available from https://doi.org/10.1097/JGP.0b013e31820119da.

Wilson, R. S., Bennett, D. A., de Leon, C. F. M., Bienias, J. L., Morris, M. C., & Evans, D. A. (2005). Distress proneness and cognitive decline in a population of older persons. *Psychoneuroendocrinology, 30*(1), 11–17. Available from https://doi.org/10.1016/j.psyneuen.2004.04.005.

Wilson, R. S., Krueger, K. R., Arnold, S. E., Schneider, J. A., Kelly, J. F., Barnes, L. L., ... Bennett, D. A. (2007). Loneliness and risk of Alzheimer disease. *Archives of General Psychiatry, 64*(2), 234–240. Available from https://doi.org/10.1001/archpsyc.64.2.234.

Wilson, R. S., Schneider, J. A., Arnold, S. E., Bienias, J. L., & Bennett, D. A. (2007). Conscientiousness and the incidence of Alzheimer disease and mild cognitive impairment. *Archives of General Psychiatry, 64*(10), 1204–1212. Available from https://doi.org/10.1001/archpsyc.64.10.1204.

Wilson, R. S., Schneider, J. A., Boyle, P. A., Arnold, S. E., Tang, Y., & Bennett, D. A. (2007). Chronic distress and incidence of mild cognitive impairment. *Neurology, 68*(24), 2085–2092. Available from https://doi.org/10.1212/01.wnl.0000264930.97061.82.

Yaffe, K., Vittinghoff, E., Lindquist, K., Barnes, D., Covinsky, K. E., Neylan, T., ... Marmar, C. (2010). Post-traumatic stress disorder and risk of dementia among U.S. post-traumatic stress disorder and risk of dementia among U.S. veterans. *Archives of General Psychiatry, 67*(6), 608–613. Available from https://doi.org/10.1001/archgenpsychiatry.2010.61.

Zubenko, G. S. (1996). Clinicopathologic and neurochemical correlates of major depression and psychosis in primary dementia. *International Psychogeriatrics, 8*(3), 219–223. Available from https://doi.org/10.1017/S1041610297003384.

Zubenko, G. S., Zubenko, W. N., McPherson, S., Spoor, E., Marin, D. B., Farlow, M. R., ... Sunderland, T. (2003). A collaborative study of the emergence and clinical features of the major depressive syndrome of Alzheimer's disease. *American Journal of Psychiatry, 160*(5), 857–866. Available from https://doi.org/10.1176/appi.ajp.160.5.857.

Zufferey, V., Donati, A., Popp, J., Meuli, R., Rossier, J., Frackowiak, R., ... Kherif, F. (2017). Neuroticism, depression, and anxiety traits exacerbate the state of cognitive impairment and hippocampal vulnerability to Alzheimer's disease. *Alzheimer's and Dementia: Diagnosis, Assessment and Disease Monitoring*, 7, 107–114. Available from https://doi.org/10.1016/j.dadm.2017.05.002.

PART II

Advanced topics in dementia research

CHAPTER 4

Alzheimer's disease in the pupil: pupillometry as a biomarker of cognitive processing in Alzheimer's disease

Mohamad El Haj[1,2,3] and Ahmed A. Moustafa[4,5,6]
[1]Laboratoire de Psychologie des Pays de la Loire (LPPL-EA 4638), Nantes Université, Univ Angers, Nantes, France
[2]Unité de Gériatrie, Centre Hospitalier de Tourcoing, Tourcoing, France
[3]Institut Universitaire de France, Paris, France
[4]School of Psychology, Western Sydney University, Sydney, NSW, Australia
[5]MARCS Institute for Brain and Behaviour, Western Sydney University, Sydney, NSW, Australia
[6]Department of Human Anatomy and Physiology, The Faculty of Health Sciences, University of Johannesburg, Johannesburg, South Africa

Introduction

Alzheimer's disease (AD) is a neurodegenerative disorder and a major public health issue that is rising at an alarming rate as the population ages. Despite all the efforts to develop effective pharmacological therapies, AD is currently incurable. In light of the lack of efficient treatment, a main focus in AD management is the identification of individuals who are at risk of AD; this is to predict disease onset and, ideally, to maintain cognitive and functional abilities before substantial neurological decline. Two main biomarkers can be used to reflect the onset of AD, namely, the presence of amyloid-β (Aβ) and tau pathology in cerebrospinal fluid (CSF) and the reduced volume of the hippocampus as can be observed on structural magnetic resonance imaging (MRI). While these biomarkers are widely used to detect AD, there is still need to identify better markers of the disease, ideally, a noninvasive (as CSF tests require medical intervention) and inexpensive test (as MRI is typically expensive). To tackle this challenge, we investigate whether pupil dilation can offer a noninvasive and inexpensive marker of cognitive decline in AD. More specifically, we offer a case study in which we evaluate whether pupil dilation can index cognitive effort in a patient with AD. By doing so, we aim to reveal whether pupil dilation has the potential to mirror cognitive processing in AD.

Alzheimer's Disease
DOI: https://doi.org/10.1016/B978-0-12-821334-6.00014-4

Our study builds on a substantial body of research demonstrating how pupil size can index cognitive effort in nonpathological population. However, before turning to this literature, it would be of interest to illustrate basic functioning of the pupil. Pupil diameter typically varies from 1.5 to 9.0 mm (Kret & Sjak-Shie, 2019; Wyatt, 1995). In standard light conditions, pupil diameter is about 3.0 mm (Kret & Sjak-Shie, 2019; Wyatt, 1995). The pupil diameter is controlled by the dilator muscles that increase the diameter of pupil and sphincter muscles that decrease the diameter (Kawasaki, 1999; Sirois & Brisson, 2014). The dilator and sphincter muscles serve to optimize vision by modulating the amount of light that reaches the retina; whereas the pupil constricts in brighter conditions, it dilates in darker conditions. This light reaction typically occurs in about 200 ms. In AD, changes in pupillary light reflex can be observed. A study has reported low pupillary light reflex in AD patients after treatment with tropicamide (Granholm et al., 2003). These changes can be associated with changes in the locus coeruleus, which is a site of accumulation of Aβ in AD (Murphy, 2019). Hence, pupil activity can mirror neuropathological changes in AD. Critically, and as illustrated below, pupil activity mirrors cognitive effort.

There is a large body of research demonstrating how pupil size increases with cognitive effort on working memory tasks, especially on span tasks. On span tasks, participants are typically invited to repeat a string of numbers in the same order (i.e., forward spans) or in reverse order (i.e., backward spans). Research has consistently demonstrated that pupil diameter increases with each digit retained in digit span tasks until the length of the digits exceeds the capacity of working memory, at which pupil diameter begins to plateau or even diminish (Alnaes et al., 2014; Cabestrero, Crespo, & Quirós, 2009; Granholm, Asarnow, Sarkin, & Dykes, 1996; Peavler, 1974; Wahn, Ferris, Hairston, & Konig, 2016). This research demonstrates that pupil dilation can be a reliable and valid psychophysiological marker of cognitive effort.

The effect of cognitive effort on pupil dilation has also been reported by research demonstrating how pupil dilation increases during processing of complex versus simple sentences (Just & Carpenter, 1993), during complex versus simple visual search (Porter, Troscianko, & Gilchrist, 2007), and during the performance of executive function tasks (Brown et al., 1999; Brown, van Steenbergen, Kedar, & Nieuwenhuis, 2014; Hasshim & Parris, 2015; Laeng, Orbo, Holmlund, & Miozzo, 2011; Steinhauer, Siegle, Condray, & Pless, 2004). Pupil dilation has been also observed during memory retrieval. A study has investigated pupil dilation during autobiographical retrieval (i.e., retrieval of personal information) by inviting participants to remember personal events while their pupil size was monitored by eye-tracking glasses (El Haj, Janssen, Gallouj, & Lenoble, 2019). In a control condition, participants were invited to count aloud. Analysis demonstrated larger pupil size during autobiographical memory retrieval than during the control task and this dilation was attributed to the cognitive load required to reconstruct the memories. A similar conclusion was drawn from a study dealing with future thinking, that is, the ability to generate hypothetical scenarios in the future. A study has recorded pupil size of participants during two conditions: during retrieval of past personal information and during construction of events that may occur in the future (El Haj & Moustafa, 2020). Results demonstrated a larger pupil size during future than past thinking. This larger pupil size was attributed to the cognitive effort as required to recombine available information into novel scenarios during future thinking. Taken together, there is a large body of research demonstrating how pupil size increases with cognitive effort.

While there is a wealth of research on the effects of cognitive effort on pupil dilation, this research was not concerned with AD. There is even, to the very best of our knowledge, no published research on whether pupil dilation can mirror cognitive processing in general. Thus we offer the first case study in which we evaluate whether pupil dilation can index cognitive effort in a patient with AD. More specifically, we measured pupil size during an effortful task (i.e., counting backwards from 100 by seven) and during a less effortful control task consisting of counting aloud from one. We expected larger pupil size during the effortful task compared with the control one. By doing so, we aimed to reveal whether pupil dilation has the potential to mirror cognitive processing in patients with AD.

Method

The case study

Mr. L is 67 years of age, has had 10 years of formal education, is right-handed, and is a native French speaker. Mr. L is living in his own home with his wife. His amnestic mild AD diagnosis was established in 2019 based on National Institute of Neurological and Communicative Disorders and Stroke and the Alzheimer's Disease and Related Disorders

Association (NINCDS-ADRDA) diagnostic criteria (McKhann et al., 2011). During the study, we assessed cognitive performances of Mr. L with the Mini-Mental State Examination (Folstein, Folstein, & McHugh, 1975) on which he obtained a score of 24/30 points.

Procedures and materials

Mr. L was tested on an effortful task and a control one. During the two tasks, Mr. L wore eye-tracking glasses and faced a white wall. The study occurred in a quiet room at the neurological department of the Hospital of Nantes. To ensure that differences in pupil dilation were not caused by differences in retinal illumination, blinds were closed and the lightness of the room (60-watt fluorescent lamp) was the same in the two conditions. Prior to the experiment, Mr. L was informed that the experiment was related to eye-tracking research and cognition in general. In order not to influence his performance, Mr. L was not provided with further details regarding pupil dilation and its relationship to cognitive processing.

In the effortful condition, we applied procedures from the Mini-Mental State Examination by inviting Mr. L to count aloud backwards from 100 by sevens (e.g., 93, 86, 79...) during 1 minute. In the control condition, we invited Mr. L to count aloud up from 1 during 1 minute. The two conditions were implemented while pupil dilation was recorded using the Pupil Capture software. Participants wore eye-tracking glasses (Pupil Lab) consisting of a remote pupil-tracking system that uses infrared illumination with 200 Hz sampling rate and a gaze position accuracy of <0.1 degrees. Prior to each past and future event, calibration was made by inviting Mr. L to fixate on a black cross (a 5×5 cm cross, printed on an A4 white paper fixated at the wall center) that was used as a calibration reference; the cross was withdrawn after calibration. During the two conditions, Mr. L was seated in front of a white wall and the distance between him and wall was approximately 30–50 cm. We invited Mr. L not to look outside the wall, but was free to explore all parts of it. The wall did not display any visual stimuli (e.g., drawings, windows). Regarding dependent variables, we calculated the mean of pupil dilation (in mm) during the two conditions.

Results

The mean pupil size of Mr. L during the effortful task and control task was, respectively, 3.3 mm and 2.7 mm. In other words, and as illustrated

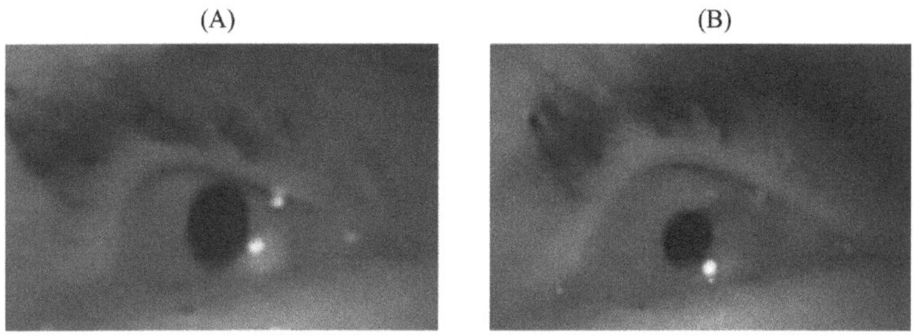

Figure 4.1 An illustration of pupil diameter of Mr. L during the effortful (A) and control (B) tasks.

in Fig. 4.1, larger pupil size was observed during the effortful task than during the control task.

Discussion

We offer the first case study about the effects of cognitive processing on pupil size in AD. With this case study, we demonstrated how a larger pupil diameter can be observed during an effortful task than during a control task. By doing this, we demonstrate how pupil size can mirror cognitive processing in AD.

The larger pupil size during the effortful task than during the control one, as observed in Mr. L, can be attributed to the cognitive load of the effortful task. This assumption can be supported by the large body of research demonstrating how pupil size increases with cognitive effort (Ahern & Beatty, 1979; Alnaes et al., 2014; Beatty, 1982; Cabestrero et al., 2009; Granholm et al., 1996; Karatekin, Couperus, & Marcus, 2004; Peavler, 1974; Piquado, Isaacowitz, & Wingfield, 2010; Wahn et al., 2016). This research supports the assumption that pupil size can indicate how intensive the cognitive load is. This cognitive load assumption fits with our procedures because counting backward from 100 by seven requires more cognitive processing (i.e., activation of the executive system in working memory to manipulate and reproduce the sequence in the reversed order) compared with the control task. The latter task requires less cognitive load as it simply involves ascribing a point value to the number "1" and so on. Taken together, the larger pupil size during the effortful task than during the control task, as observed in Mr. L, can be attributed to the cognitive load of the effortful task.

Besides the cognitive load account, pupil size as observed in Mr. L can be attributed to activities in the sympathetic (i.e., adrenergic) and parasympathetic (i.e., cholinergic) autonomic nervous systems. Pupil dilation involves inhibitory regulation of the Edinger–Westphal nucleus as well as activation of preganglionic sympathetic neurons (Kawasaki, 1999). Thus the larger pupil size during the effortful task than during the control task, as observed in Mr. L, can be attributed to higher inhibitory activity in the Edinger–Westphal nucleus as well as to higher activation of preganglionic sympathetic neurons, resulting in decreased activity of sphincter muscles and decreased activity of dilator muscles during the effortful task. Thus the variations in pupil size across our two experimental conditions can mirror both adrenergic and cholinergic activities. This issue is important because AD has been associated with deficits in both adrenergic (Kelly et al., 2017; Prettyman, Bitsios, & Szabadi, 1997) and cholinergic activities (Ornek, Dag, & Ornek, 2015; Scinto et al., 2001; Shen & Wu, 2015; Singh & Verma, 2020). Pupillometry can be used as a biomarker of both adrenergic and cholinergic activities in AD. Critically, a study has found significant correlations between pupil size and Aβ and tau levels in CSF in AD (Frost et al., 2013). Hence, pupillometry can mirror the neuropathology of AD and, as observed in our case study, cognitive processing. We believe that our study provides evidence on the value of pupillometry as a potential biomarker of cognitive processing in AD as little research has evaluated the value of pupillometry, and retinal activity in general, as a potential biomarker in AD.

We are of course aware that the case study design limits the generalization of our findings. Although case studies are useful for generating innovative research, the experimental design of case studies do not have the level of methodological rigor to support the findings. Thus the generalization of our findings should be viewed with caution. That said, we believe that our case study paves the way to the design and use of pupillometry as a practical and noninvasive biomarker of cognitive processing in AD. This biomarker may significantly contribute to precision medicine that seeks to implement key breakthrough technological and scientific advances, especially in neurosciences, to offer individually tailored diagnosis (Hampel, Vergallo, Perry, Lista, & Alzheimer Precision Medicine, 2019).

To summarize, the pupil size offers substantial information on cognitive and neurological processing. As proposed by our study, pupil size can offer a valuable biomarker of cognitive processing in AD. Our study case paves the innovative research on the use of pupillometry as a practical and noninvasive biomarker of cognitive processing in AD, leading, in light of the lack of efficient treatment, to better diagnosis of AD.

References

Ahern, S., & Beatty, J. (1979). Pupillary responses during information processing vary with Scholastic Aptitude Test scores. *Science (New York, N.Y.), 205*(4412), 1289–1292.

Alnaes, D., Sneve, M. H., Espeseth, T., Endestad, T., van de Pavert, S. H., & Laeng, B. (2014). Pupil size signals mental effort deployed during multiple object tracking and predicts brain activity in the dorsal attention network and the locus coeruleus. *Journal of Vision, 14*(4), 1–6. Available from https://doi.org/10.1167/14.4.1.

Beatty, J. (1982). Task-evoked pupillary responses, processing load, and the structure of processing resources. *Psychological Bulletin, 91*(2), 276–292.

Brown, G. G., Kindermann, S. S., Siegle, G. J., Granholm, E., Wong, E. C., & Buxton, R. B. (1999). Brain activation and pupil response during covert performance of the Stroop Color Word task. *Journal of the International Neuropsychological Society: JINS, 5* (4), 308–319.

Brown, S. B., van Steenbergen, H., Kedar, T., & Nieuwenhuis, S. (2014). Effects of arousal on cognitive control: Empirical tests of the conflict-modulated Hebbian-learning hypothesis. *Frontiers in Human Neuroscience, 8*, 23. Available from https://doi.org/10.3389/fnhum.2014.00023.

Cabestrero, R., Crespo, A., & Quirós, P. (2009). Pupillary dilation as an index of task demands. *Perceptual and Motor Skills, 109*(3), 664–678. Available from https://doi.org/10.2466/pms.109.3.664-678.

El Haj, M., Janssen, S. M. J., Gallouj, K., & Lenoble, Q. (2019). Autobiographical memory iIncreases pupil dilation. *Translational Neuroscience, 10*, 280–287. Available from https://doi.org/10.1515/tnsci-2019-0044.

El Haj, M., & Moustafa, A. A. (2020). Pupil dilation as an indicator of future thinking. *Neurological Sciences: Official Journal of the Italian Neurological Society and of the Italian Society of Clinical Neurophysiology*. Available from https://doi.org/10.1007/s10072-020-04533-z.

Folstein, M. F., Folstein, S. E., & McHugh, P. R. (1975). Mini-mental state. A practical method for grading the cognitive state of patients for the clinician. *Journal of Psychiatric Research, 12*(3), 189–198.

Frost, S., Kanagasingam, Y., Sohrabi, H. R., Taddei, K., Bateman, R., Morris, J., & Martins, R. N. (2013). Pupil response biomarkers distinguish amyloid precursor protein mutation carriers from non-carriers. *Current Alzheimer Research, 10*(8), 790–796. Available from https://doi.org/10.2174/15672050113109990154.

Granholm, E., Asarnow, R. F., Sarkin, A. J., & Dykes, K. L. (1996). Pupillary responses index cognitive resource limitations. *Psychophysiology, 33*(4), 457–461.

Granholm, E., Morris, S., Galasko, D., Shults, C., Rogers, E., & Vukov, B. (2003). Tropicamide effects on pupil size and pupillary light reflexes in Alzheimer's and Parkinson's disease. *International Journal of Psychophysiology: Official Journal of the International Organization of Psychophysiology, 47*(2), 95–115. Available from https://doi.org/10.1016/s0167-8760(02)00122-8.

Hampel, H., Vergallo, A., Perry, G., Lista, S., & Alzheimer Precision Medicine, I. (2019). The Alzheimer Precision Medicine Initiative. *Journal of Alzheimer's Disease: JAD, 68* (1), 1–24. Available from https://doi.org/10.3233/JAD-181121.

Hasshim, N., & Parris, B. A. (2015). Assessing stimulus-stimulus (semantic) conflict in the Stroop task using saccadic two-to-one color response mapping and preresponse pupillary measures. *Attention Perception Psychophysics*, 77(8), 2601–2610. Available from https://doi.org/10.3758/s13414-015-0971-9.

Just, M. A., & Carpenter, P. A. (1993). The intensity dimension of thought: Pupillometric indices of sentence processing. *Canadian Journal of Experimental Psychology/Revue Canadienne de Psychologie Experimentale*, *47*(2), 310–339.

Karatekin, C., Couperus, J. W., & Marcus, D. J. (2004). Attention allocation in the dual-task paradigm as measured through behavioral and psychophysiological responses. *Psychophysiology*, *41*(2), 175–185. Available from https://doi.org/10.1111/j.1469-8986.2004.00147.x.

Kawasaki, A. (1999). Physiology, assessment, and disorders of the pupil. *Current Opinion in Ophthalmology*, *10*(6), 394–400.

Kelly, S. C., He, B., Perez, S. E., Ginsberg, S. D., Mufson, E. J., & Counts, S. E. (2017). Locus coeruleus cellular and molecular pathology during the progression of Alzheimer's disease. *Acta Neuropathologica Communications*, *5*(1), 8. Available from https://doi.org/10.1186/s40478-017-0411-2.

Kret, M. E., & Sjak-Shie, E. E. (2019). Preprocessing pupil size data: Guidelines and code. *Behavior Research Methods*, *51*(3), 1336–1342. Available from https://doi.org/10.3758/s13428-018-1075-y.

Laeng, B., Orbo, M., Holmlund, T., & Miozzo, M. (2011). Pupillary Stroop effects. *Cognitive Processing*, *12*(1), 13–21. Available from https://doi.org/10.1007/s10339-010-0370-z.

McKhann, G., Knopman, D. S., Chertkow, H., Hyman, B. T., Jack, C. R., Jr., Kawas, C. H., & Phelps, C. H. (2011). The diagnosis of dementia due to Alzheimer's disease: Recommendations from the National Institute on Aging-Alzheimer's Association workgroups on diagnostic guidelines for Alzheimer's disease. *Alzheimer's & Dementia: The Journal of the Alzheimer's Association*, 7(3), 263–269. Available from https://doi.org/10.1016/j.jalz.2011.03.005.

Murphy, C. (2019). Olfactory and other sensory impairments in Alzheimer disease. *Nature Reviews Neurology*, *15*(1), 11–24. Available from https://doi.org/10.1038/s41582-018-0097-5.

Ornek, N., Dag, E., & Ornek, K. (2015). Corneal sensitivity and tear function in neurodegenerative diseases. *Current Eye Research*, *40*(4), 423–428. Available from https://doi.org/10.3109/02713683.2014.930154.

Peavler, W. S. (1974). Pupil size, information overload, and performance differences. *Psychophysiology*, *11*(5), 559–566.

Piquado, T., Isaacowitz, D., & Wingfield, A. (2010). Pupillometry as a measure of cognitive effort in younger and older adults. *Psychophysiology*, *47*(3), 560–569. Available from https://doi.org/10.1111/j.1469-8986.2009.00947.x.

Porter, G., Troscianko, T., & Gilchrist, I. D. (2007). Effort during visual search and counting: Insights from pupillometry. *Quarterly Journal of Experimental Psychology (Hove)*, *60*(2), 211–229. Available from https://doi.org/10.1080/17470210600673818.

Prettyman, R., Bitsios, P., & Szabadi, E. (1997). Altered pupillary size and darkness and light reflexes in Alzheimer's disease. *Journal of Neurology, Neurosurgery, and Psychiatry*, *62*(6), 665–668. Available from https://doi.org/10.1136/jnnp.62.6.665.

Scinto, L. F., Frosch, M., Wu, C. K., Daffner, K. R., Gedi, N., & Geula, C. (2001). Selective cell loss in Edinger-Westphal in asymptomatic elders and Alzheimer's patients. *Neurobiology of Aging*, *22*(5), 729–736. Available from https://doi.org/10.1016/s0197-4580(01)00235-4.

Shen, J., & Wu, J. (2015). Chapter Ten—Nicotinic cholinergic mechanisms in Alzheimer's disease. In M. De Biasi (Ed.), *International Review of Neurobiology* (Vol. 124, pp. 275–292). Academic Press.
Singh, A., & Verma, S. (2020). Use of ocular biomarkers as a potential tool for early diagnosis of Alzheimer's disease. *Indian Journal of Ophthalmology*, *68*(4), 555–561. Available from https://doi.org/10.4103/ijo.IJO_999_19.
Sirois, S., & Brisson, J. (2014). Pupillometry. *Wiley Interdisciplinary Review of Cognitive Science*, *5*(6), 679–692. Available from https://doi.org/10.1002/wcs.1323.
Steinhauer, S. R., Siegle, G. J., Condray, R., & Pless, M. (2004). Sympathetic and parasympathetic innervation of pupillary dilation during sustained processing. *International Journal of Psychophysiology: Official Journal of the International Organization of Psychophysiology*, *52*(1), 77–86. Available from https://doi.org/10.1016/j.ijpsycho.2003.12.005.
Wahn, B., Ferris, D. P., Hairston, W. D., & Konig, P. (2016). Pupil sizes scale with attentional load and task experience in a multiple object tracking task. *PLoS One*, *11*(12), e0168087. Available from https://doi.org/10.1371/journal.pone.0168087.
Wyatt, H. J. (1995). The form of the human pupil. *Vision Research*, *35*(14), 2021–2036.

CHAPTER 5

Cognitive and neural correlates of vitamin D deficiency: focus on healthy aging and Alzheimer's disease

Ahmed A. Moustafa[1,2,3], Wafa Jaroudi[4] and Abdrabo Soliman[5]
[1]School of Psychology, Western Sydney University, Sydney, NSW, Australia
[2]MARCS Institute for Brain and Behaviour, Western Sydney University, Sydney, NSW, Australia
[3]Department of Human Anatomy and Physiology, The Faculty of Health Sciences, University of Johannesburg, Johannesburg, South Africa
[4]School of Social Sciences and Psychology, Western Sydney University, Sydney, NSW, Australia
[5]Department of Social Sciences, College of Arts and Sciences, Qatar University, Doha, Qatar

Introduction

Vitamin D deficiency is a global public health concern due to its influence on health and well-being in which individuals have less than the minimum required concentration of vitamin D in their body and brain. Holick and Chen (2008) have estimated that 1 billion individuals worldwide have vitamin D deficiency. Although the definitive levels of vitamin D have not been agreed upon worldwide, Pettersen (2016) defined levels of vitamin D as follows: insufficient as being levels less than 50 nmol/L, low sufficient being between 50 and 75 nmol/L, high sufficient being between 75 and 100 nmol/L, and supratherapeutic being equal to or more than 100 nmol/L (Pettersen, 2016, p. 467).

This literature review considers many epidemiological, nutrition-oriented, and psychological studies that investigate the role of serum 25-hydroxyvitamin D [25(OH)D], including studies that investigate serum 25(OH)D and vitamin D hormone (VDH). In addition, our review covers several studies that have investigated the relationship between vitamin D and cognitive functions. Some studies just mention that only global cognitive function in general is affected (Annweiler, Fantino, Le Gall, & Beauchet, 2011; Bartali, Devore, Grodstein, & Kang, 2014), while others suggest that specific functions are affected by low vitamin D levels including verbal fluency (Breitling et al., 2012; Przybelski & Binkley, 2007), episodic memory (Feart et al., 2017), as well as experiences of

Alzheimer's Disease
DOI: https://doi.org/10.1016/B978-0-12-821334-6.00008-9

confusion and general memory problems (Annweiler et al., 2011; Bartali et al., 2014; Breitling et al., 2012; Brouwer-Brolsma, de Groot, & van de Rest, 2015; Brouwer-Brolsma et al., 2014; Feart et al., 2017; Przybelski & Binkley, 2007; Van der Schaft et al., 2013; Vieira, Mendy, Prado, Gasana, & Albatineh, 2015). Annweiler et al (2011) explored the association between vitamin D and global cognition in a sample of 462 elderly French aged 65 years and older. After administering several cognitive tests (e.g., Clock drawing test) and grouping participants into cognitively normal and impaired, the results revealed that participants with worse global cognition had lower vitamin D levels. However, these results were not specific to certain cognitive aspects (Annweiler et al., 2011). Interestingly, a recent Korean study has shown that vitamin D levels are not strongly associated with cognitive function in a sample of older adults (Lee et al., 2020).

In addition, some studies have investigated the relationship between vitamin D levels and brain functioning (Hua et al., 2012; McCann & Ames, 2008; Minasyan, Keisala, Lou, Kalueff, & Tuohimaa, 2007; Schneider et al., 2014; Turner, Young, McGrath, Eyles, & Burne, 2013), memory (Becker, Eyles, McGrath, & Grecksch, 2005; Larsen, 2010; Pettersen, 2017), and verbal fluency (Pettersen, 2016, 2017). Most of these studies have shown that vitamin D deficiency does indeed play a role in cognitive and neural processes (Sultan et al., 2020), especially hippocampus-based cognitive processes (Croll et al., 2020). It is possible that reduced levels of vitamin D alter hippocampus function, subsequently leading to Alzheimer's disease (AD) (Apostolova et al., 2006; Dhikav & Anand, 2011; Hyman, Van Hoesen, Damasio, & Barnes, 1984; Jaroudi et al., 2017).

Individuals who have vitamin D deficiency were found to show faster cognitive decline over the years especially in functions such as verbal episodic memory. In contrast to individuals with vitamin D deficiency, older adults who have adequate (more than 50 nmol/L) levels of vitamin D are suggested to have less risk of developing dementia, and demonstrate better cognitive preservation (Feart et al., 2017). In terms of vitamin D affecting the structure of the hippocampus, a review study of rodent studies reveal that a lack of vitamin D can affect the hippocampal structure and function leading to impaired cognitive functions such as latent inhibition not only during early stages of life but also in later age (Lardner, 2015).

Some studies have focused on investigating the role played by vitamin D levels in healthy aging (Bartali et al., 2014; Feart et al., 2017; Larsen, 2010), and people in mid to late life (Schneider et al., 2014; Vieira et al., 2015). For example, in a longitudinal study, Bartali et al. (2014) investigated the relationship between plasma levels of vitamin D and cognitive functioning in 1185 elderly females aged between 60 and 70 years. In this study, participants' plasma levels were measured in previous years between 1989 and 1990 with a follow-up cognitive evaluation through a telephone interview in 1999 and cognitive assessments. Although the measures of participants' plasma levels of vitamin D were conducted approximately 9 years before evaluation, the findings of the study revealed that those with higher levels of vitamin D had better cognitive function suggesting that the effects of vitamin D levels on cognition can last for a long time. Therefore vitamin D levels does not only have beneficial effects in the short term (Pettersen, 2017), but also in the long term, which could protect against age-related cognitive decline (Bartali et al., 2014).

Other studies have investigated the relationship between vitamin D levels and brain disorders including dementia and AD (Annweiler et al., 2011, 2013; Feart et al., 2017; Kandiah, Jion, & Ng, 2011; Llewellyn, Lang, Langa, & Melzer, 2010; Wong et al., 2017). Kandiah et al. (2011) highlight the role of vitamin D in relation to mild cognitive impairment (MCI) and AD. By comparing cognitively normal and cognitively impaired (MCI and AD) participants' cognition using assessments such as the mini mental state examination (MMSE) and Alzheimer's Disease Assessment Scale—Cognitive Subscale (ADAS-Cog), Kandiah et al. (2011) found a direct association between vitamin D and cognitive performance. Overall, cognitively normal individuals had larger levels vitamin D than the cognitively impaired individuals. By conducting a comparison within the cognitively impaired group, Kandiah et al. (2011) revealed that individuals who had worse vitamin D levels also had worse MMSE and ADAS-Cog scores [for similar findings, also see Aguilar-Navarro et al. (2019) and Jeon, Jung, & Kim (2020)]. Such findings suggest that vitamin D levels do impact cognition and, can contribute to the development of cognitive impairment and diseases such as AD.

Given the strong links between vitamin D levels and cognitive process even before the development of neurological diseases (e.g., AD), some studies have investigated the impact of vitamin D supplements on the brain and cognition. For example, Pettersen (2017) investigated the influence of high-dose vitamin D3 supplementation on cognitive function among young adults with insufficient vitamin D (<75 nmol/L). Findings of the study highlight that nonverbal (visual) memory is ameliorated by higher doses of vitamin D supplementation. Verbal memory and other cognitive domains on the other hand were not influenced by vitamin D levels. This study suggests that vitamin D status is an independent predictor of immediate visuospatial memory, which is also associated with executive functioning processes (Pettersen, 2017; Zugic Soares et al., 2019).

Further, because of the existing links between vitamin D deficiency and neurological disorders, several studies have investigated the therapeutic effect of vitamin D supplements. For example, Pettersen (2017) ran an 18-month long study investigating whether high doses of vitamin D supplements enhanced cognition in a total of 82 healthy individuals whose vitamin D levels at baseline did not significantly differ. In this study, some participants were given high doses of vitamin D supplements, while other participants were given low doses. The participants' cognition was evaluated using several tests focusing on functions such as verbal fluency and digit span. The results show that serum 25(OH)D levels are significantly higher in participants who received high doses of vitamin D in comparison to participants who received low doses. Functionally, unlike a low dose, the high vitamin D supplement dose has led to improved nonverbal (visual) memory. Thus it is suggested that memory is improved after taking high doses of vitamin D in comparison to low doses, even over short periods of times such as 18 months.

Annweiler et al. (2012) investigated the long-term effect of vitamin D3 supplementation on cognition (especially executive function) in older adults with memory problems. Participants in this study were outpatients from a memory clinic that were prescribed oral vitamin D3 in the form of 25OHD at an approximate dose of 34 ng/mL. To examine the relationship between vitamin D3 supplements and cognition, Annweiler et al. (2012) assessed global cognition using measures such as the MMSE and Cognitive Assessment Battery (CAB), as well as the Frontal Assessment Battery (FAB). Using multiple regression to analyze the relationship between pre and postvitamin D3 levels and cognition, at 16-month follow-up, participants' vitamin D3 was significantly increased. Moreover, vitamin D3 supplementation has led to a higher (i.e., improved) MMSE, CAB, and FAB scores when compared to individuals in the control group who did not receive any vitamin D3 supplementation. Further, increased calcifediol (25OHD), a VDH produced in the liver, was noted to be associated with FAB scores of participants receiving vitamin D3 supplementation.

Along the lines of investigating the relationship of vitamin D doses and its benefits to cognition, another study found that a response to vitamin D supplements is also related to genetic differences among individuals, perhaps suggesting why only some individuals benefit more from vitamin D supplements (Waterhouse et al., 2014). To evaluate the relationship between vitamin D and genetic variability in 350 older adult participants over 12 months, Waterhouse et al. (2014) used blood samples to measure vitamin D levels, DNA samples to analyze single-nucleotide polymorphisms (SNPs) genetic variation, and questionnaires to ask about lifestyle (e.g., diet, outdoor exposure). There were two groups who received different doses of vitamin D3: 30,000 IU dose and 60,000 IU dose. Data analysis revealed responsiveness to vitamin D3 was related to Body Mass Index and SNPs associated with the regulation of vitamin D metabolism, sun sensitivity, as well as skin, eye, and hair pigmentation. Waterhouse et al. (2014) also found an increase of 1.8 nmol/L in the 60,000 IU group for every 100 IU vitamin D intake each day; and an increase of 2.2 nmol/L in the 30,000 IU group for every 100 IU vitamin D intake each day. It was further found that responsiveness of vitamin D3 supplementation decreased as the dose increased. Explaining participants' responsiveness to vitamin D3 increase and dosage, it is suggested that the vitamin D3 could become more saturated when converted to 25(OH)D. Regarding genetic variability, the alleles rs10766197 and rs10741657 found in the gene CYP2R1 were more responsive in the hydroxylation of vitamin D3 to serum 25(OH)D than participants with other alleles of the same gene. On the other hand, the allele rs2282679 found in the gene GC/DBP was associated to less responsiveness to vitamin D3. As such, Waterhouse et al. (2014) provide an explanation for why some individuals may be more responsive and benefit more from different doses of vitamin D supplementation and possibly why some individuals may be more genetically prone to being vitamin D deficient [for similar points, see Wang et al. (2020)].

This literature review is divided into the following subsections: (1) cognitive function and vitamin D in animals and (2) vitamin D levels in healthy aging and AD. Throughout the review, we will discuss the impact of vitamin D supplements, including diets that increase vitamin D, which have positive impacts on the brain and cognition.

Animal studies on cognitive function and vitamin D levels

A variety of studies on vitamin D have been conducted using rats (Becker et al., 2005; Erbas et al., 2014; Turner et al., 2013) and mice (Brouwer-Brolsma et al., 2014; Erbas et al., 2014; Minasyan et al., 2007). Findings from Becker et al. (2005) included adult rats with low vitamin D levels at the prenatal stage showing significantly impaired latent inhibition, which is considered a biomarker of hippocampus function (Grecksch, Bernstein, Becker, Hollt, & Bogerts, 1999; Moustafa, Myers, & Gluck, 2009; Schmajuk, Larrauri, & Labar, 2007). This could suggest that adult rats who had low vitamin D levels before birth had slower learning abilities in comparison to adult rats who had healthy levels of vitamin D in their prenatal stages (Becker et al., 2005). Such findings therefore highlight that not only do vitamin D levels impact cognition in individuals later in old age (Bartali et al., 2014) but also before birth (Becker et al., 2005).

Several studies suggest that vitamin D is a vital component both for brain development and adult brain functioning. Minasyan et al. (2007) explored this idea in vitamin D receptor (VDR) mutant male adult mice aged between 3 and 5 months. They have assessed working memory function using the Y-maze test where mice relied on their spatial memory to complete a maze in 30-minute interval trials. After 2 weeks, the mice were food-deprived for 24 hours where they were then required to complete another Y-maze task with 2-hour reinforcement using food. Conducting statistical analyses, the study found VDR mice to have unchanged spatial and working memory assessed by comparing performance on the two maze tasks completed. However, in terms of emotional behavior and motor deficit, emotional/anxiety states were observed but not depression. Therefore it could be suggested that although VDR affects emotional states such as experiencing anxiety, it does not change some aspects of memory (Minasyan et al., 2007). The results of Minasyan et al. (2007) differ from the results of previous studies (Becker et al., 2005; Przybelski & Binkley, 2007) that emphasized an association between vitamin D and cognitive function, leading to a gap in agreement between research on vitamin D and cognition.

Animal research suggests VDH is important for cognitive functioning. Hua et al. (2012) conducted an animal study on 46 male rats (Sprague-Dawley breed) to investigate what level of VDH and progesterone doses were most effective for the treatment of brain damage: low (1 μg/kg, VDH), medium (2.5 μg/kg, VDH), and high (5 μg/kg, VDH). The researchers first damaged the medial frontal cortex (MFC) of the rats. After surgically impairing the MFC by 10 days, the rats completed the water maze task in which the time spent outside the maze was measured as a representation of the rats' behavioral responsiveness, short-term memory, and long-term memory. Following behavioral and learning assessment, the rats' brains were later analyzed to examine the degeneration of neurons and brain tissue. It was found that progesterone alone did not influence cognition or preservation of spatial memory but, a combined treatment using VDH and progesterone did. Specifically, low and medium doses of VDH and progesterone were most effective in preserving cognition and, the low dose was effective in improving learning. Behaviorally, it was noted that progesterone and low VDH improved rats' exhibition of anxiety as suggested by observations of rats' reduced tendency to be close to the walls of the water maze. In terms of long-term memory, it was found that the low VDH and progesterone dose was most effective. Therefore it could be suggested that VDH is effective at certain doses to improve/preserve cognition in animals with brain damage. These findings might be translatable to humans who experience cognitive impairment where vitamin D can probably reverse or at least improve their memory decline. Further, combining both progesterone and vitamin/hormone D could be even more beneficial if taken in the right doses (low to medium dose in rats). Similarly, another study found that combining vitamin D with resveratrol may protect against the development of dementia and AD in a mouse model of AD (Cheng et al., 2017). To our knowledge, similar studies have not been done in human participants. Recently, one study found that increasing vitamin D levels enhanced hippocampal function and improved memory performance in rat model of AD (Mehri et al., 2020).

Other studies have investigated the effect of memantine (a common anti-AD drug) in combination with vitamin D on cognition. In one study, patients treated with memantine show similar cognitive performance to patients treated with memantine and vitamin D supplementation. That is, memantine alone improves AD patients' cognition while vitamin D supplementation does not add any further significant cognitive improvements (Wong et al., 2017). As such, it could be understood that vitamin D could be important before the development of cognitive diseases such as AD. However, after developing AD, it may then be too late to improve cognition with only vitamin D supplementation where instead, common anti-AD drug interventions may be more beneficial in improving cognition. Therefore it is important to advocate the early intervention of vitamin D supplementation in avoidance of cognitive issues, AD, and associated AD drug interventions. This suggests that unlike resveratrol and progesterone, memantine may enhance cognitive performance when combined with vitamin D supplements. Other studies also found that vitamin D supplement do beta-amyloids in AD patients (Jia et al., 2019).

Among studies that have assessed vitamin D status and cognitive function, Turner et al. (2013) investigated the effects of developmental vitamin D deficiency on attention measured using the 5-choice serial reaction time task (5C-SRT) and the 5-choice continuous performance test. By measuring visuospatial attention in a control group of rats and rats whose mothers had vitamin D deficiency during pregnancy, Turner et al. (2013) investigated the effects of low vitamin D levels on sustained attention and vigilance. Results show that low levels of vitamin D during pregnancy were associated with impaired attention. In line with the idea of low vitamin D during pregnancy affecting offspring later in life, Becker et al. (2005) investigated the impact of prenatal vitamin D deficiency on latent inhibition (which is a measure of hippocampus function). Becker et al. (2005) found that prenatal vitamin D depletion was linked with subtle and discrete alterations in learning and memory in rats. Therefore it could be suggested that during pregnancy, not having the sufficient vitamin D levels required can affect unborn babies' cognition in terms of learning and memory, highlighting the importance of vitamin D even before birth and not only over time as we age.

Recently, some studies have investigated how exactly vitamin D supplements can be neuroprotective against the development of AD. In one study, Yamini, Ray, and Chopra (2018) investigated the impact of pretreatment with vitamin D3 supplements on several neural and behavioral measures in rats. The results show vitamin D3 improved spatial memory (which is associated with hippocampal function), improved the function of acetylcholine, and protected against death of neurons in the hippocampus and cortex. This suggests that vitamin D3 can protect people against the development of AD.

Healthy aging, dementia, and Alzheimer's disease

Many previous studies have investigated the association of vitamin D with dementia, AD, and healthy aging (Bartali et al., 2014; Brouwer-Brolsma et al., 2014; Ertilav, Barcin, & Ozdem, 2020; Eymundsdottir et al., 2020; Feart et al., 2017; Kalra, Teixeira, & Diniz, 2020; Kandiah et al., 2011; Llewellyn et al., 2010; Prabhakar et al., 2015; SanMartín, Ponce, Salech, & Behrens, 2015; Schneider et al., 2014; Wong et al., 2017; Zhao et al., 2020), but for conflicting findings, see Duchaine et al. (2020). In this section, we will first discuss existing studies on healthy aging followed by studies on AD.

Breitling et al. (2012) have investigated the relationship between vitamin D3 and several cognitive functions in individuals aged 65 years and older. Specifically, they have measured vitamin D levels and cognitive functions at baseline and 5-year follow-up. Cognitive functions were assessed using the cognition telephone interview (COGTEL) while vitamin D levels were measured using chemiluminescence methods. By comparing participants' patterns of vitamin D and cognition, women with lower vitamin D levels had worse cognitive functions than women who had higher vitamin D levels as measured by the COGTEL, and vitamin D samples especially on functions of verbal fluency and inductive reasoning. A similar relationship was also found between male participants where vitamin D levels determined better or worse cognitive performance overall (Breitling et al., 2012).

Przybelski and Binkley (2007) have explored the effect of serum 25 (OH)D and vitamin B12 on cognition in 80 older adult individuals. To measure cognition, the researchers used the MMSE. The researchers also sampled the participants' concentrations of 25(OH)D and vitamin B12 using blood samples within 6 hours of measuring cognition. After conducting regression analysis, Przybelski and Binkley (2007) found that MMSE scores to be positively correlated with serum 25(OH)D concentrations but not with serum vitamin B12 concentrations. As such, it is suggested that vitamin D concentrations can affect cognition (also see Shih, Lee, Hsu, Wang, & Fuh, 2020).

In a recent study, Pettersen (2016) assessed the impact of vitamin D on executive functioning in 142 healthy adults. Executive function was assessed using verbal fluency, digit span backward, Cambridge Neuropsychological Test Automated Battery (CANTAB) Spatial Working Memory, and One Touch Stockings of Cambridge. The results show that the levels of vitamin D are a significant independent predictor of verbal fluency. Moreover, further statistical tests involved spline analysis, which specified the relationship between vitamin D and executive function to be positive and near linear when regarding verbal fluency and 25(OH)D levels of 100 nmol/L and greater (Pettersen, 2016, p. 467). Therefore it could be suggested that 100 nmol/L of vitamin D concentrations are possibly an optimal level for executive functions such as verbal fluency (Pettersen, 2016).

Another population-based epidemiological study investigated the association between plasma 25(OH)D concentrations and cognition in participants aged between 45 and 60 years (Assmann et al., 2015). The study examined the relationship between cognitive performance and education level using a variety of neuropsychological tests, including phonemic and semantic fluency tasks (e.g., Trail Making Test, forward and backward digit span) to evaluate verbal memory and short-term/working memory. This study reported a relationship between 25(OH)D concentrations and the cognitive function of "short-term/working memory" in participants who had low education. This was especially evident on the backward digit span test. On the other hand, in participants with more educational attainment, plasma vitamin D levels were only associated with phonemic fluency. Therefore it is suggested that higher plasma vitamin D is more associated to individuals' cognitive functions than their level of education. Moreover in terms of age, the study found that, although on average, younger individuals performed better on cognitive function than older individuals and adequate plasma vitamin D levels were more favorable in terms of indicating higher performance on cognitive function in comparison to individuals with low plasma vitamin D levels.

Similar to previously mentioned studies (Breitling et al., 2012; Prabhakar et al., 2015), Larsen (2010) examined the relationship between vitamin D intake and cognitive performance. This study focused on the relationship between the diets of 1916 elderly individuals (aged between 70 years and 74 years) and their cognitive function. At baseline, 64% of the participants were low on vitamin D measured at a level of 10 μg/d or more being the required amount. After taking cod liver oil supplements, Larsen (2010) found that vitamin D intake has increased, meeting approximately 76% of the required amount. The conclusion drawn from such findings is that fish (or fish-related supplements) assists and contributes to the benefit of vitamin D absorption by approximately 38% of the participants. Not only did fish (or fish-related supplements) increase vitamin D levels, but it also had cognitive effects on participants. That is, increased vitamin D levels improved participants' scores on verbal fluency (measured using the Semantic Fluency Task) and episodic memory (measured using the Kendrick Object Learning Test). Therefore it is suggested that lean fish such as cod is beneficial in the improvement of vitamin D levels, which consequently shows improvement in participants' cognitive functions.

The mechanism by which consuming fish can assist with slowing cognitive decline includes reducing inflammation and oxidative stress (Qin et al., 2014), which can be very harmful to the brain, affecting its functioning. Fish improves cognition, especially in functions of verbal, immediate, and delayed memory (Qin et al., 2014). Moreover, fish supplementation usually contains omega-3 fatty acids, eicosapentaenoic acid, and docosahexaenoic acid, which reduce inflammation and oxidative stress in the central nervous system (Qin et al., 2014). As such, fish supplements support and improve cognition (O'Connor, Power, Fitzgerald, & O'Toole, 2012). Not only do fish supplements have cognitive benefit but also fish in its own form. This is supported by a study by Nurk et al. (2007) finding that unprocessed fatty and lean fish improve cognition.

Prabhakar et al. (2015) have investigated the impact of vitamin D in a sample of 140 individuals aged 60 years and older. The authors hypothesized that vitamin D sufficiency is protective against the development of vascular dementia (VaD). Measures of vitamin D levels revealed that individuals with low vitamin D (<12 ng/mL) had an increased chance of developing VaD. Importantly, the study also reported that a deficiency of vitamin D and hypertension was associated with even higher risk of VaD than either alone (Prabhakar et al., 2015). Moreover, one recent study found that vitamin D level did not significantly between different subtypes of dementia patients, including AD, dementia with Lewy body, VaD, and frontotemporal dementia (Soysal, Dokuzlar, Erken, Dost Gunay, & Isik, 2020).

Many studies have reported a relationship between vitamin D and cognition (e.g., Assmann et al., 2015; Schneider et al., 2014). In a population of 1652 participants aged between 45 and 65 years old and of mixed color/race (Caucasian and Black), the association between 25(OH)D concentrations and cognition was studied over several years(Schneider et al., 2014). Reasons for the sample being mixed color/race were that individuals of darker skin would generally have lower levels of vitamin D than fair-skinned individuals. Schneider et al. (2014) have measured the levels of 25(OH)D using serum samples, as well calcium, phosphorus, and parathyroid hormone. The authors also measured cognitive processing, using the delayed word recall test (which tests verbal learning and recent memory), the digit symbol substitution test (which measures executive function and processing speed), and the word fluency test (which measures executive function and language). The results revealed no association between low 25(OH)D levels and cognition. However, Schneider et al. (2014) reported that a nonsignificant finding of this relationship may be related to the age of participants as they suggest low 25(OH)D levels are more common in individuals aged 60 years and older.

Vieira et al. (2015) have evaluated the association of vitamin D status with cognitive skills in people with type II diabetes, pre-diabetes, and undiagnosed diabetes. Data from 37,973 people with an average age of 47

($\pm$17) years was analyzed to investigate how diabetes can contribute to physical limitations, memory, and vitamin A, E, and D levels. The study also concluded that early-onset diagnosis of diabetes in individuals before the age of 40 was associated with memory problems frequently found in individuals specifically with low vitamin D levels but not A or E. Therefore it is suggested that early-onset diabetes and low vitamin D contribute to cognitive problems such as confusion and memory impairment, as well as limited physical abilities (e.g., inability to stand for long periods of time) (Vieira et al., 2015). According to findings of Vieira et al. (2015), it is suggested that individuals who develop diabetes in their early midlife stages and are low in vitamin D may face greater cognitive consequences later in life in comparison to individuals who develop diabetes later in life and have good levels of vitamin D. Cognitive consequences may be linked to that of cognitive decline, dementia, and also AD, as the development of diabetes is linked to vitamin D deficiency leading to cognitive problems (Vieira et al., 2015), and vitamin D deficiency is also related to the development of dementia and AD (Chai et al., 2019; Littlejohns et al., 2014), but for conflicting results on the relationship between vitamin D deficiency and the development of AD, see Yang, Chen, Li, and Zhou (2019).

Annweiler et al. (2011) studied the relationship between serum 25OHD and episodic memory and capacity planning in 462 elderly individuals aged 65 years and older. Participants involved in the study were split into two groups (impaired and normal cognition) based on global cognitive measures of three-word list recall tests and the clock drawing test. By comparing the two groups' performance scores, results show that worse global cognitive performance is associated with low 25(OH)D concentrations overall but, not specific to episodic memory, capacity planning, or spatial arrangement.

In a population-based observational study of 1639 elderly participants aged 65 years and older, Breitling et al. (2012) used the COGTEL to assess participants' cognitive function over 5 years. This measure is very similar to the MMSE as it also screens for the diagnoses of dementia diseases such as AD. Furthermore, standardized questionnaires were completed by participants if additional information regarding their cognitive function was needed. Breitling et al. (2012) found that low vitamin D was associated with worse cognitive function. Unlike the results of Breitling et al. (2012), Walker (2010) provided a review of five articles published between 2000 and 2010 showing disagreement regarding the relationship between cognitive dysfunction and vitamin D in elderly individuals. Along these lines, Luckhaus et al. (2009) did not find a relationship between vitamin D and cognition. In a sample of 47 participants (19 with MCI, 20 with AD, and 8 cognitively normal) who had their vitamin D levels measured using biochemical tests, cognition was tested using measures such as the MMSE and the Clinical Dementia Rating Scale. Statistical comparison of participants' biochemical and cognitive data revealed no significant difference between the MCI, AD, and cognitively normal participants' cognition and vitamin D levels.

Bartali et al. (2014) took a more general approach to investigate the relationship between vitamin D levels and cognitive functions in 1185 women aged between 60 and 70 years. This longitudinal study ran for 9 years in which participants had their plasma 25(OH)D level biomarkers measured in 1989–90, followed up by a telephone interview to assess their cognition in 1999 along with other cognitive assessments (e.g., telephone adaptation of the MMSE, East Boston Memory test) between 1999 and the early 2000s. After the assessment of vitamin D using radioimmunoassay, the study found that participants with lower plasma 25(OH)D levels had worse cognitive functions when compared with those with normal levels, even after 9 years at follow-up. One important finding of this study is that in the sixth year of the study, the vitamin D status of participants was not significantly associated with cognitive decline. However, in relation to this specific finding, Assmann et al. (2015) suggest this could be due to age differences among the population in which age can have some slight effect on the results. As such, a longitudinal follow-up could show differences in findings, as shown at the ninth year follow-up by Bartali et al. (2014).

Feart et al. (2017) performed a study similar to that of Bartali et al. (2014) across 12 years among a nondemented population of 916 individuals aged 65 years and older. To assess participants' cognition, several tests were administered such as the MMSE and the Benton Visual Retention test. On the other hand, to measure plasma 25(OH)D concentrations, participants blood samples were collected throughout the years and analyzed to assess for vitamin D levels. Results show that individuals with vitamin D deficiency have a faster progression of cognitive decline and an increased risk of developing AD over the 12 years of the study. On the other hand, participants with vitamin D at sufficient levels were associated with better cognition, suggesting a slower progression of cognitive decline (if any) over age and time of the study and decreasing the likelihood of developing AD. As such, it could be concluded that adequate intake and maintenance of vitamin D, especially in older adults, is imperative to slow down the progression of cognitive decline and dementia (Feart et al., 2017), which prevent the development of AD.

Littlejohns et al. (2014) explored the relationship between vitamin D and dementia and AD in old-aged adults using a longitudinal study running for approximately 5.6 years. To measure vitamin D, serum 25(OH) D samples were collected; to screen for and measure dementia and AD cognitive assessments, various screenings such as magnetic resonance imaging, questionnaires, and reviewing of cognitive performance history among other procedures were used. Littlejohns et al. (2014) found participants who developed dementia and AD were vitamin D deficient or severely deficient. Participants who had vitamin D deficiency were 51% more likely to develop dementia while those severely deficient were 121% more likely to develop dementia. These results were similarly found for developing AD. Authors suggest that vitamin D becomes a risk for developing dementia and AD when levels are below 50 nmol/L. Several studies have suggested that vitamin D deficiency is linked to cognitive problems (Bartali et al., 2014; Feart et al., 2017), and Littlejohns et al. (2014) also suggest that a lack of vitamin D may lead to impaired cognition, increasing the chances of dementia (but for different results, see Nourhashemi et al., 2018). In another more recent longitudinal study (for around 6 years) on more than 1700 older adults without dementia, Zhao et al. (2020) found that high vitamin D is associated with reduced likelihood to develop dementia.

Conclusion

The critical importance of vitamin D in public health, especially in aging and AD, has been explained in the studies reviewed above. It is apparent that adequate vitamin D levels is important for a variety of health reasons, mainly improving and preserving cognitive functioning and reducing risk factors associated with the development of dementia and AD, as reported in several animal and human studies (e.g., Feart et al., 2017; Przybelski & Binkley, 2007). Further, it is important to note that providing healthy older adults and individuals with MCI with vitamin D supplements may protect against the development of AD, as argued by SanMartin et al. (2018).

We have also highlighted that not only does vitamin D have an effect later in age, but also before birth and during the early stages of life in terms of learning and, memory (e.g., Becker et al., 2005). Importantly, our review shows that vitamin D deficiency does occur in old age and even before the development of AD (e.g., Feart et al., 2017). According to extensive literature search and review, and to the best of our knowledge, there is an indication that vitamin D is a lot more important before AD development as vitamin D supplementation can have a less effect (if not minimal) on cognition after the development of AD.

Future implications

To further explore the relationship between vitamin D and cognition in healthy and healthy aging (as well as dementia and AD), it is important to review explorations of gender ratios related to vitamin D deficiency. Review of gender ratios of vitamin D deficiency, dementia, and AD may provide better understanding of which gender is more prone to vitamin D deficiency, making them more susceptible to brain diseases (dementia and AD). It is also important to understand the relationship and effects of vitamin D and cognition in young adults. Doing so will assist targeting factors contributing to later cognitive disease developments and promote implementations to improve cognition and vitamin D. Such exploration is imperative as, to the best of our knowledge, there is a clear gap in assessments of vitamin D status in young populations (adolescents and infants), which can have detrimental effects on cognition. Importantly, as in animal studies, research on older adults and patients with AD should investigate the combined effects of vitamin D along with resveratrol (Cheng et al., 2017) or progesterone (Hua et al., 2012) on the brain and cognition.

Ethics approval and consent to participate

No ethics applications were required for this review article.

Consent for publication

All authors have consented to the publication of this study.

Availability of data and material

Our study does not include any data or material, as it is a review article.

Competing interests

We do not have any competing interests.

Funding

We did not obtain any funding for this study.

Authors' contributions

AS wrote the first draft. Subsequently, AAM and WJ have worked on several drafts of this review study. All ideas presented here were suggested by AAM, WJ, and AS.

References

Aguilar-Navarro, S. G., Mimenza-Alvarado, A. J., Jimenez-Castillo, G. A., Bracho-Vela, L. A., Yeverino-Castro, S. G., & Avila-Funes, J. A. (2019). Association of vitamin D with mild cognitive impairment and Alzheimer's dementia in older Mexican adults. *Revista de Investigacion Clinica; Organo del Hospital de Enfermedades de la Nutricion*, *71*(6), 381–386. Available from https://doi.org/10.24875/RIC.19003079.

Annweiler, C., Fantino, B., Le Gall, D., & Beauchet, O. (2011). Which cognitive function is influenced by vitamin D among older high-functioning community-dwellers? *Alzheimer's & Dementia*, 7(4), S590–S591.

Annweiler, C., Montero-Odasso, M., Llewellyn, D. J., Richard-Devantoy, S., Duque, G., & Beauchet, O. (2013). Meta-analysis of memory and executive dysfunctions in relation to vitamin D. *Journal of Alzheimer's Disease*, *37*(1), 147–171.

Annweiler, C., Rolland, Y., Schott, A. M., Blain, H., Vellas, B., Herrmann, F. R., et al. (2012). Higher vitamin D dietary intake is associated with lower risk of Alzheimer's disease: a 7-year follow-up. The Journals of Gerontology. Series A, Biological Sciences and Medical Sciences, 67, 1205–1211. https://doi.org/10.1093/gerona/gls107.

Apostolova, L. G., Dutton, R. A., Dinov, I. D., Hayashi, K. M., Toga, A. W., Cummings, J. L., & Thompson, P. M. (2006). Conversion of mild cognitive impairment to Alzheimer disease predicted by hippocampal atrophy maps. *Archives of Neurology*, *63*(5), 693–699.

Assmann, K. E., Touvier, M., Andreeva, V. A., Deschasaux, M., Constans, T., Hercberg, S., & Kesse-Guyot, E. (2015). Midlife plasma vitamin D concentrations and performance in different cognitive domains assessed 13 years later. *The British Journal of Nutrition*, *113*(10), 1628–1637. Available from https://doi.org/10.1017/S0007114515001051.

Bartali, B., Devore, E., Grodstein, F., & Kang, J. H. (2014). Plasma vitamin D levels and cognitive function in aging women: The Nurses' Health Study. *The Journal of Nutrition, Health & Aging*, *18*(4), 400–406.

Becker, A., Eyles, D. W., McGrath, J. J., & Grecksch, G. (2005). Transient prenatal vitamin D deficiency is associated with subtle alterations in learning and memory functions in adult rats. *Behavioural Brain Research*, *161*(2), 306–312. Available from https://doi.org/10.1016/j.bbr.2005.02.015.

Breitling, L. P., Perna, L., Muller, H., Raum, E., Kliegel, M., & Brenner, H. (2012). Vitamin D and cognitive functioning in the elderly population in Germany. *Experimental Gerontology*, *47*(1), 122–127. Available from https://doi.org/10.1016/j.exger.2011.11.004.

Brouwer-Brolsma, E. M., de Groot, L. C. P. G. M., & van de Rest, O. (2015). Chapter 63 - Vitamin D and the association with cognitive performance, cognitive decline, and dementia. In Diet and Nutrition in Dementia and Cognitive Decline, pp. 679–700.

Brouwer-Brolsma, E. M., Schuurman, T., de Groot, L. C., Feskens, E. J., Lute, C., Naninck, E. F., & Steegenga, W. T. (2014). No role for vitamin D or a moderate fat diet in aging induced cognitive decline and emotional reactivity in C57BL/6 mice. *Behavioural Brain Research*, *267*, 133–143. Available from https://doi.org/10.1016/j.bbr.2014.03.038.

Chai, B., Gao, F., Wu, R., Dong, T., Gu, C., Lin, Q., & Zhang, Y. (2019). Vitamin D deficiency as a risk factor for dementia and Alzheimer's disease: An updated meta-analysis. *BMC Neurology*, *19*(1), 284. Available from https://doi.org/10.1186/s12883-019-1500-6.

Cheng, J., Rui, Y., Qin, L., Xu, J., Han, S., Yuan, L., & Wan, Z. (2017). Vitamin D combined with resveratrol prevents cognitive decline in SAMP8 mice. *Current Alzheimer Research*, *14*(8), 820–833. Available from https://doi.org/10.2174/1567205014666170207093455.

Croll, P. H., Boelens, M., Vernooij, M. W., van de Rest, O., Zillikens, M. C., Ikram, M. A., & Voortman, T. (2020). Associations of vitamin D deficiency with MRI markers of brain health in a community sample. *Clinical Nutrition (Edinburgh, Scotland)*, *40* (1), 72–78. Available from https://doi.org/10.1016/j.clnu.2020.04.027.

Dhikav, V., & Anand, K. (2011). Potential predictors of hippocampal atrophy in Alzheimer's disease. *Drugs & Aging*, *28*(1), 1–11. Available from https://doi.org/10.2165/11586390-000000000-00000, 1 [pii].

Duchaine, C. S., Talbot, D., Nafti, M., Giguere, Y., Dodin, S., Tourigny, A., & Laurin, D. (2020). Vitamin D status, cognitive decline and incident dementia: The Canadian study of health and aging. *Canadian Journal of Public Health. Revue Canadienne de Sante Publique*, *111*(3), 312–321. Available from https://doi.org/10.17269/s41997-019-00290-5.

Erbas, O., Solmaz, V., Aksoy, D., Yavasoglu, A., Sagcan, M., & Taskiran, D. (2014). Cholecalciferol (vitamin D 3) improves cognitive dysfunction and reduces inflammation in a rat fatty liver model of metabolic syndrome. *Life Sciences*, *103*(2), 68–72. Available from https://doi.org/10.1016/j.lfs.2014.03.035.

Ertilav, E., Barcin, N. E., & Ozdem, S. (2020). Comparison of serum free and bioavailable 25-hydroxyvitamin D levels in Alzheimer's disease and healthy control patients. Laboratory Medicine. https://doi.org/10.1093/labmed/lmaa066

Eymundsdottir, H., Chang, M., Geirsdottir, O. G., Gudmundsson, L. S., Jonsson, P. V., Gudnason, V., & Ramel, A. (2020). Lifestyle and 25-hydroxy-vitamin D among community-dwelling old adults with dementia, mild cognitive impairment, or normal cognitive function. *Aging Clinical and Experimental Research*, *32*(12), 2649–2656. Available from https://doi.org/10.1007/s40520-020-01531-1.

Feart, C., Helmer, C., Merle, B., Herrmann, F. R., Annweiler, C., Dartigues, J. F., & Samieri, C. (2017). Associations of lower vitamin D concentrations with cognitive decline and long-term risk of dementia and Alzheimer's disease in older adults. *Alzheimer's & Dementia*, *13*(11), 1207–1216. Available from https://doi.org/10.1016/j.jalz.2017.03.003.

Grecksch, G., Bernstein, H. G., Becker, A., Hollt, V., & Bogerts, B. (1999). Disruption of latent inhibition in rats with postnatal hippocampal lesions. *Neuropsychopharmacology: Official Publication of the American College of Neuropsychopharmacology*, *20*(6), 525–532.

Holick, M. F., & Chen, T. C. (2008). Vitamin D deficiency: A worldwide problem with health consequences. *The American Journal of Clinical Nutrition*, *87*(4), 1080S–1086S.

Hua, F., Reiss, J. I., Tang, H., Wang, J., Fowler, X., Sayeed, I., & Stein, D. G. (2012). Progesterone and low-dose vitamin D hormone treatment enhances sparing of memory following traumatic brain injury. *Hormones and Behavior*, *61*(4), 642–651. Available from https://doi.org/10.1016/j.yhbeh.2012.02.017.

Hyman, B. T., Van Hoesen, G. W., Damasio, A. R., & Barnes, C. L. (1984). Alzheimer's disease: Cell-specific pathology isolates the hippocampal formation. *Science (New York, N.Y.)*, *225*(4667), 1168–1170.

Jaroudi, W., Garami, J., Garrido, S., Hornberger, M., Keri, S., & Moustafa, A. A. (2017). Factors underlying cognitive decline in old age and Alzheimer's disease: The role of the hippocampus. *Reviews in the Neurosciences*, *28*(7), 705–714. Available from https://doi.org/10.1515/revneuro-2016-0086.

Jeon, Y. J., Jung, S. J., & Kim, H. C. (2020). Does serum vitamin D level affect the association between cardiovascular health and cognition? Results of the Cardiovascular and Metabolic Diseases Etiology Research Center (CMERC) study. *European Journal of Neurology: The Official Journal of the European Federation of Neurological Societies*, *28*(1), 48–55. Available from https://doi.org/10.1111/ene.14496.

Jia, J., Hu, J., Huo, X., Miao, R., Zhang, Y., & Ma, F. (2019). Effects of vitamin D supplementation on cognitive function and blood Abeta-related biomarkers in older adults with Alzheimer's disease: A randomised, double-blind, placebo-controlled trial. *Journal of Neurology, Neurosurgery, and Psychiatry*, *90*(12), 1347–1352. Available from https://doi.org/10.1136/jnnp-2018-320199.

Kalra, A., Teixeira, A. L., & Diniz, B. S. (2020). Association of vitamin D levels with incident all-cause dementia in longitudinal observational studies: A systematic review and meta-analysis. *The Journal of Prevention of Alzheimer's Disease*, 7(1), 14–20. Available from https://doi.org/10.14283/jpad.2019.44.

Kandiah, N., Jion, Y., & Ng, A. (2011). Severity of cognitive impairment is correlated to levels of vitamin D. *Alzheimer's & Dementia*, 7(4), S612–S613.

Lardner, A. L. (2015). Vitamin D and hippocampal development-the story so far. *Frontiers in Molecular Neuroscience*, *8*, 58. Available from https://doi.org/10.3389/fnmol.2015.00058.

Larsen, L. (2010). *Intake of vitamin D in relation to cognition in the elderly*. Master's thesis, University of Oslo.

Lee, D. H., Chon, J., Kim, Y., Seo, Y. K., Park, E. J., Won, C. W., & Soh, Y. (2020). Association between vitamin D deficiency and cognitive function in the elderly Korean population: A Korean frailty and aging cohort study. *Medicine (Baltimore)*, *99* (8), e19293. Available from https://doi.org/10.1097/MD.0000000000019293.

Littlejohns, T. J., Henley, W. E., Lang, I. A., Annweiler, C., Beauchet, O., Chaves, P. H., & Llewellyn, D. J. (2014). Vitamin D and the risk of dementia and Alzheimer disease. *Neurology*, *83*(10), 920–928. Available from https://doi.org/10.1212/WNL.0000000000000755.

Llewellyn, D. J., Lang, I. A., Langa, K. M., & Melzer, D. (2010). Vitamin D and cognitive impairment in NHANES III. *Alzheimer's & Dementia*, *6*(4), S81.

Luckhaus, C., Mahabadi, B., Grass-Kapanke, B. et al. (2009). Blood biomarkers of osteoporosis in mild cognitive impairment and Alzheimer's disease. *Journal of Neural Transmission*, 116(7), 905–911.

McCann, J. C., & Ames, B. N. (2008). Is there convincing biological or behavioral evidence linking vitamin D deficiency to brain dysfunction? *The FASEB Journal*, *22*(4), 982–1001.

Mehri, N., Haddadi, R., Ganji, M., Shahidi, S., Soleimani Asl, S., Taheri Azandariani, M., & Ranjbar, A. (2020). Effects of vitamin D in an animal model of Alzheimer's disease: Behavioral assessment with biochemical investigation of hippocampus and serum. *Metabolic Brain Disease*, *35*(2), 263–274. Available from https://doi.org/10.1007/s11011-019-00529-7.

Minasyan, A., Keisala, T., Lou, Y. R., Kalueff, A. V., & Tuohimaa, P. (2007). Neophobia, sensory and cognitive functions, and hedonic responses in vitamin D receptor mutant mice. *The Journal of Steroid Biochemistry and Molecular Biology*, *104*(3-5), 274–280. Available from https://doi.org/10.1016/j.jsbmb.2007.03.032.

Moustafa, A. A., Myers, C. E., & Gluck, M. A. (2009). A neurocomputational model of classical conditioning phenomena: A putative role for the hippocampal region in associative learning. *Brain Research*, *1276*, 180–195.

Nourhashemi, F., Hooper, C., Cantet, C., Feart, C., Gennero, I., Payoux, P., ... Multidomain Alzheimer Preventive Trial/Data Sharing Alzheimer (DSA) study group. (2018). Cross-sectional associations of plasma vitamin D with cerebral beta-amyloid in older adults at risk of dementia. *Alzheimer's Research & Therapy*, *10*(1), 43. Available from https://doi.org/10.1186/s13195-018-0371-1.

Nurk, E., Drevon, C. A., Refsum, H., Solvoll, K., Vollset, S. E., Nygard, O., & Smith, A. D. (2007). Cognitive performance among the elderly and dietary fish intake: The Hordaland Health Study. *The American Journal of Clinical Nutrition*, *86*(5), 1470–1478. Available from https://doi.org/10.1093/ajcn/86.5.1470.

O'Connor, E. M., Power, S. E., Fitzgerald, G. F., & O'Toole, P. W. (2012). Fish-oil consumption is inversely correlated with depression and cognition decline in healthy irish elderly adults. *Proceedings of the Nutrition Society*, *71*(1).

Pettersen, J. A. (2016). Vitamin D and executive functioning: Are higher levels better? *Journal of Clinical and Experimental Neuropsychology*, *38*(4), 467–477. Available from https://doi.org/10.1080/13803395.2015.1125452.

Pettersen, J. A. (2017). Does high dose vitamin D supplementation enhance cognition?: A randomized trial in healthy adults. *Experimental Gerontology*, *90*, 90–97. Available from https://doi.org/10.1016/j.exger.2017.01.019.

Prabhakar, P., Chandra, S. R., Supriya, M., Issac, T. G., Prasad, C., & Christopher, R. (2015). Vitamin D status and vascular dementia due to cerebral small vessel disease in the elderly Asian Indian population. *Journal of the Neurological Sciences*, *359*(1-2), 108–111. Available from https://doi.org/10.1016/j.jns.2015.10.050.

Przybelski, R. J., & Binkley, N. C. (2007). Is vitamin D important for preserving cognition? A positive correlation of serum 25-hydroxyvitamin D concentration with cognitive function. *Archives of Biochemistry and Biophysics*, *460*(2), 202–205. Available from https://doi.org/10.1016/j.abb.2006.12.018.

Qin, B., Plassman, B. L., Edwards, L. J., Popkin, B. M., Adair, L. S., & Mendez, M. A. (2014). Fish intake is associated with slower cognitive decline in Chinese older adults. *The Journal of Nutrition*, *144*(10), 1579–1585. Available from https://doi.org/10.3945/jn.114.193854.

SanMartín, C., Ponce, D., Salech, F., & Behrens, M. (2015). Correction of vitamin D status improves lymphocyte susceptibility to oxidative cell death in mild cognitive impairtment patients. *Journal of the Neurological Sciences*, *357*, e137–e138.

SanMartin, C. D., Henriquez, M., Chacon, C., Ponce, D. P., Salech, F., Rogers, N. K., & Behrens, M. I. (2018). Vitamin D increases Abeta140 plasma levels and protects lymphocytes from oxidative death in mild cognitive impairment patients. *Current Alzheimer Research*, *15*(6), 561–569. Available from https://doi.org/10.2174/1567205015666171227154636.

Schmajuk., Larrauri., & Labar. (2007). Reinstatement of conditioned fear and the hippocampus: An attentional-associative model. *Behavioural Brain Research*, *177*(2), 242–253.

Schneider, A. L., Lutsey, P. L., Alonso, A., Gottesman, R. F., Sharrett, A. R., Carson, K. A., & Michos, E. D. (2014). Vitamin D and cognitive function and dementia risk in a biracial cohort: The ARIC Brain MRI Study. *European Journal of Neurology*, *21*(9), 1211–1218. Available from https://doi.org/10.1111/ene.12460, e1269-1270.

Shih, E. J., Lee, W. J., Hsu, J. L., Wang, S. J., & Fuh, J. L. (2020). Effect of vitamin D on cognitive function and white matter hyperintensity in patients with mild Alzheimer"s disease. *Geriatrics & Gerontology International*, *20*(1), 52–58. Available from https://doi.org/10.1111/ggi.13821.

Soysal, P., Dokuzlar, O., Erken, N., Dost Gunay, F. S., & Isik, A. T. (2020). The relationship between dementia subtypes and nutritional parameters in older adults. *Journal of the American Medical Directors Association, 21*(10), 1430–1435. Available from https://doi.org/10.1016/j.jamda.2020.06.051.

Sultan, S., Taimuri, U., Basnan, S. A., Ai-Orabi, W. K., Awadallah, A., Almowald, F., & Hazazi, A. (2020). Low vitamin D and its association with cognitive impairment and dementia. *Journal of Aging Research, 2020*, 6097820. Available from https://doi.org/10.1155/2020/6097820.

Turner, K. M., Young, J. W., McGrath, J. J., Eyles, D. W., & Burne, T. H. (2013). Cognitive performance and response inhibition in developmentally vitamin D (DVD)-deficient rats. *Behavioural Brain Research, 242*, 47–53. Available from https://doi.org/10.1016/j.bbr.2012.12.029.

Van der Schaft, J., Koek, H. L., Dijkstra, E., Verhaar, H. J., Van der Schouw, Y. T., & Emmelot-Vonk, M. H. (2013). The association between vitamin D and cognition: A systematic review. *Ageing Research Reviews, 12*(4), 1013–1023. Available from https://doi.org/10.1016/j.arr.2013.05.004.

Vieira, E. R., Mendy, A., Prado, C. M., Gasana, J., & Albatineh, A. N. (2015). Falls, physical limitations, confusion and memory problems in people with type II diabetes, undiagnosed diabetes and prediabetes, and the influence of vitamins A, D and E. *Journal of Diabetes and Its Complications, 29*(8), 1159–1164. Available from https://doi.org/10.1016/j.jdiacomp.2015.08.005.

Walker, E. (2010). *The relationship between vitamin D deficiency and cognitive decline in the geriatric population*. Paper 194, MSc thesis, Pacific University.

Wang, L., Qiao, Y., Zhang, H., Zhang, Y., Hua, J., Jin, S., & Liu, G. (2020). Circulating vitamin D levels and Alzheimer's disease: A Mendelian randomization study in the IGAP and UK biobank. *Journal of Alzheimer's Disease: JAD, 73*(2), 609–618. Available from https://doi.org/10.3233/JAD-190713.

Waterhouse, M., Tran, B., Armstrong, B. K., Baxter, C., Ebeling, P. R., English, D. R., & Neale, R. E. (2014). Environmental, personal, and genetic determinants of response to vitamin D supplementation in older adults. *The Journal of Clinical Endocrinology and Metabolism, 99*(7), E1332–E1340. Available from https://doi.org/10.1210/jc.2013-4101.

Wong, D., Bellyou, M., Beauchet, O., Montero-Odasso, M., Annweiler, C., & Bartha, R. (2017). Combined memantine and vitamin d treatment provides the same cognitive benefit as memantine alone in a chronically vitamin D deficient double-transgenic mouse model of Alzheimer's disease. *Alzheimer's & Dementia, 13*(7), P269–P270.

Yamini, P., Ray, R. S., & Chopra, K. (2018). Vitamin D3 attenuates cognitive deficits and neuroinflammatory responses in ICV-STZ induced sporadic Alzheimer's disease. *Inflammopharmacology, 26*(1), 39–55. Available from https://doi.org/10.1007/s10787-017-0372-x.

Yang, K., Chen, J., Li, X., & Zhou, Y. (2019). Vitamin D concentration and risk of Alzheimer disease: A meta-analysis of prospective cohort studies. *Medicine (Baltimore), 98*(35), e16804. Available from https://doi.org/10.1097/MD.0000000000016804.

Zhao, C., Tsapanou, A., Manly, J., Schupf, N., Brickman, A. M., & Gu, Y. (2020). Vitamin D intake is associated with dementia risk in the Washington Heights-Inwood Columbia Aging Project (WHICAP). *Alzheimer's & Dementia*. Available from https://doi.org/10.1002/alz.12096.

Zugic Soares, J., Pettersen, R., Saltyte Benth, J., Knapskog, A. B., Selbaek, G., & Bogdanovic, N. (2019). Higher vitamin D levels are associated with better attentional functions: Data from the NorCog Register. *The Journal of Nutrition, Health & Aging, 23*(8), 725–731. Available from https://doi.org/10.1007/s12603-019-1220-z.

CHAPTER 6

The effect of Alzheimer's disease on the thalamus

Rasu Karki[1] and Ahmed A. Moustafa[1,2,3]
[1]School of Psychology, Western Sydney University, Sydney, NSW, Australia
[2]MARCS Institute for Brain and Behaviour, Western Sydney University, Sydney, NSW, Australia
[3]Department of Human Anatomy and Physiology, The Faculty of Health Sciences, University of Johannesburg, Johannesburg, South Africa

Introduction

The thalamus is a large mass of gray matter in the forebrain with two major components, the dorsal and the ventral thalamus (Murray Sherman, 2006). The dorsal part of the thalamus consists of a set of relay nuclei that transmit primary sensory information to cerebral cortex (Murray Sherman, 2006). Early in development, the thalamus gets divided into two progenitor domains, the caudal and the rostral domain. The caudal domain is responsible for the development of glutamatergic neurons in different thalamic nuclei (Blumenfeld & Gummadavelli, 2018). Similarly, the rostral domain also leads to the formation of thalamic reticular nucleus.

The major thalamic nuclei are divided into relay nuclei, association nuclei, midline/intralaminar nuclei, and the reticular nucleus (Blumenfeld & Gummadavelli, 2018; Li, Lopez-Heurta et al., 2020; Moustafa, Mcmullan, Rostron, Hewedi, & Haladjian, 2017). They act as specific relay for incoming sensory information (Gazzaniga, Ivry, & Mangun, 1998; Pelzer, Pauls, Braun, Tittgemeyer, & Timmermann, 2020; Sheridan & Tadi, 2020). The relay nuclei act as a relay station for motor and sensory information in the central nervous system and include the lateral and medial geniculate (Blumenfeld & Gummadavelli, 2018). The lateral geniculate receives sensory input from the ganglia retina and sends output to the primary visual cortex (Gazzaniga et al., 1998). However, the medial geniculate receives input from the inner ear and sends axons to primary auditory complex (Gazzaniga et al., 1998).

The association nuclei have major connection with the limbic system and are responsible for regulating sensory information. In addition, the association nuclei are also important for high-level cognitive functions (Blumenfeld & Gummadavelli, 2018). The intralaminar nuclei consist of

Alzheimer's Disease
DOI: https://doi.org/10.1016/B978-0-12-821334-6.00005-3

the dorsal and ventral groups. The dorsal group includes the paratenial and paraventricular nuclei and the ventral midline of the thalamus includes the nucleus reuniens and rhomboidal nucleus (Blumenfeld & Gummadavelli, 2018; Cassel et al., 2013). They form cortico-thalamo-cortical pathways and are involved in memory and attentional processes (De Medeiros Silva et al., 2014). Lastly, the reticular nucleus forms a shell surrounding the thalamic nuclei. The reticular nucleus regulates the processing of information between the thalamus and the cortex (Blumenfeld & Gummadavelli, 2018; Jagirdar & Chin, 2019).

The limbic nuclei of the thalamus/limbic thalamus consist of the medial-dorsal nucleus, the lateral dorsal nucleus, the anterior nucleus. and midline nuclei (Taber, Wen, Khan, & Hurley, 2015; Vertes, Linley, & Hoover, 2015). These nuclei have close connections with several cortical and subcortical areas (Taber et al., 2015). These nuclei are also important for the regulation of circadian rhythms and regulating sleep (Taber et al., 2015). The thalamus is essential for attentional focus and contributes to information processing, memory, emotional, motivational, and associative cognitive processes (De Jong et al., 2008; Fama & Edith, 2015; Gazzaniga et al., 1998; Koyama, Molfese, Milham, Mencl, & Pugh, 2020; Li, Xia et al., 2020). The thalamus is also involved in regulating consciousness and sleep (Iglesias et al., 2018). In addition, the anterior, medial-dorsal, intralaminar, and midline nuclei in the thalamus are important for memory functions (De Jong et al., 2008; Štillová et al., 2015).

Alzheimer's disease (AD) is a neurogenerative disorder associated with memory decline, impairment in language, and executive function. It is the most common form of dementia affecting more than 50 million people all over the world (Cutsuridis & Moustafa, 2014; Hodson, 2018). AD is characterized by continuous loss of pyramidal neurons within the cerebral cortex and hippocampus (Minter, Taylor, & Crack, 2015). AD results from the formation of intracellular neurofibrillary tangles and accumulation of extracellular amyloid β peptide (Aβ) plaques in the brain (Aisen et al., 2017; De Jong et al., 2008; Xu, Chen, Li, Xing, & Lu, 2019). This leads to neurodegeneration, which results in loss of neurons along with impairment in neuronal functions and synaptic and neuronal loss (Aisen et al., 2017; Lane, Hardy, & Scott, 2018). In addition, structural imaging studies have shown that the formation of neurofibrillary tangles and amyloid plaques leads to hippocampal, amygdala, medial temporal lobe, precuneus, and global gray matter atrophy (De Jong et al., 2008).

Various amyloid imaging studies in patients with AD have concluded that amyloid deposition occurs in the thalamus (Ryan et al., 2013). Several studies have also found that the cholinergic system is impacted in

AD (Hampel et al., 2018; Kanel et al., 2020). Importantly, some studies have shown that the cholinergic system plays a key role in thalamic processes (Huerta-Ocampo Hacioglu-Bay, Dautan, & Mena-Segovia, 2020; Noftz, Beebe, Mellott, & Schofield, 2020; Pita-Almenar, Yu, Lu, & Beierlein, 2014). However, several studies have shown cholinergic projection to the thalamus is not associated with AD (Kotagal, Muller, Kaufer, Koeppe, & Bohnen, 2012; Mega, 2000).

The effects of Alzheimer's disease on the thalamus

In this section, we discuss how AD may impact the thalamus, including volume reduction and neural damage to several thalamic regions.

Volume reduction

AD results from an increased presence of neurofibrillary tangles and amyloid plaques in the limbic nuclei of thalamus, along with synaptic loss (De Jong et al., 2008; Zarei et al., 2010). According to Zarei et al. (2010), the anterodorsal, centromedial, and pulvinar nuclei are the main sites of degeneration in the thalamus. The decrease in thalamic volume depicts an early sign related with degrading cognitive performance along with volume loss in gray and white matter (Zidan et al., 2019).

There is a significant decrease in the volume of thalamus in patients with AD (De Jong et al., 2008). It was found that the decrease in the volume of the thalamus correlates with impaired cognitive performance in elderly subjects when controlled for age, gender, intracranial volume, and neocortical gray matter volume (De Jong et al., 2008). Zidan et al. (2019) suggested that volume loss in the thalamus is the result of amyloid accumulation and axon degeneration in patients with AD. In addition, it was found that the anterior and dorsal parts of the thalamus are significantly smaller in AD patients than in healthy people. However, the decrease in the size of thalamus cannot be accounted to the reduction of the anterior nuclei alone. This indicates that other parts of thalamus must also undergo atrophy (De Jong et al., 2008). These neurofibrillary tangles and amyloid plaques were found in all limbic nuclei of the thalamus with greater severity of tangles in the medial-dorsal nucleus.

Moreover, Xuereb et al. (1991) found that all thalamic nuclei except the lateral posterior and central lateral nuclei show some volume reductions in AD even though these reductions did not individually show statistical significance. They found that only reticular and dorsal nucleus showed significant volume reduction. In addition, Low et al. (2019) found that the left posterior ventrolateral and ventromedial thalamus of the patients with AD showed more atrophy than other brain regions. The left ventral thalamus plays a role in verbal fluency, and damage to this brain regions may explain problems faced by patients with AD such as impaired speech, reduced verbal output, among other speech problems (Low et al., 2019). The decrease in reticular nucleus volume might reflect atrophy of the entire thalamus since the reticular nucleus consists of an outer lamina of nerve cells that form a shell around the core of thalamic nuclei. If the core shows some atrophy, then the surrounding reticular nucleus would shrink inward thus explaining the atrophy of the thalamus.

Nerve cell loss

The anterior nuclear group of the thalamus is subdivided into anteroventral, anteromedial, and anterodorsal nuclei. The anterodorsal nuclei is the major site of neurofibrillary degeneration and neuronal loss in the anterior nuclear group, which shows 80% of neuronal cell loss (Aggleton, Pralus, Nelson, & Hornberger, 2016; Xuereb et al., 1991; Zarei et al., 2010). Xuereb et al. (1991) concluded that the anteroventral and anteromedial nuclei of anterior nuclear group and the lateral dorsal nucleus do not show nerve cell loss. In addition, centromedial nucleus and pulvinar nuclei also undergo nerve cell loss in AD (Zarei et al., 2010). The anatomical/chemical basis for severe nerve cell loss in the anterodorsal nucleus and the behavioral consequences of cell loss is completely confirmed. In addition, Xuereb et al. (1991) showed that central medial nucleus also undergoes neuronal loss, but this is not associated with significant neurofibrillary tangle formation unlike the anterodorsal nucleus.

Contrary to the findings of Xuereb et al. (1991), Aggleton et al. (2016) concluded that the anteroventral thalamic nucleus of the thalamus also undergoes cell loss in AD. Moreover, according to Xuereb et al. (1991), the centromedian nucleus of the thalamus is major site of neural degeneration. The nerve cells that remain degenerated get reduced to irregular, dark, and shrunken structures, which lie within the neuropil in the centromedial nucleus. This is associated with an increased number of astrocytes, which indicates antemortem neural damage (Xuereb et al., 1991). However, Xuereb et al. (1991) found the neural degeneration to be equally severe in elderly controls and AD groups thus suggesting this degeneration cannot solely be attributed to AD. This is confirmed by the finding that the centromedian nucleus lacked neurofibrillary tangles. However, it was found that surrounding nerve cells in mediodorsal and the ventral posteromedial nucleus were not affected by AD (Xuereb et al., 1991).

It was also found that thalamus's shape changes due to the development of AD. Since the anterodorsal nucleus curves around the anteroventral nucleus and is only a small portion of the thalamic volume, it is suggested that the internal medullary lamina undergoes degeneration (Zarei et al., 2010).

The observed change in the volume of thalamus cannot be accounted for by loss of nerve cell since anterior nuclei group is only a small portion of the entire thalamic volume (Zarei et al., 2010). Iglesias et al. (2018) have hypothesized that the change in volume may be due to loss of cell nerve processes or changes in the cell glial population.

Thalamus's link to other parts of the brain

In this section, we discuss how disruptions to areas connected to the thalamus have also been implicated in the development of AD, including the hippocampus, Papez circuit, the retrosplenial cortex, retrosplenial cortex, thalamocortical network, and prefrontal-striatal loops.

Hippocampus

The loss of episodic memory is among early signs of the development of AD. The hippocampus is considered an important component for regulating episodic memory (Aggleton et al., 2016).

The hippocampus connects directly with the anterior thalamus through the fornix and to the pulvinar through the temporopulvinar tract. The hippocampus also indirectly connects with the anterior thalamus through the mammillary bodies. This hippocampus-anterior thalamus pathway is essential for episodic memory (Zarei et al., 2010). The hippocampal inputs reach the thalamus through the fornix. However, fornix shows cell loss and neurofibrillary tangle formation and atrophies in AD leading to disconnection from the hippocampal inputs. This leads to disruption in episodic memory (Aggleton et al., 2016). The source of these hippocampal inputs is the subiculum, which shows cell loss and neurofibrillary degeneration in AD. This results in disconnection between the left hippocampus and the anterior thalamic nuclei, which leads to episodic memory impairment (Aggleton & Brown, 1999; Feng et al., 2019). Along these lines, Jenkins,

Amin, Brown, and Aggleton (2006) found that hippocampal lesions decrease the gene activity in the anterior thalamic nuclei.

Papez circuit

The Papez circuit is connected to the limbic thalamus, and this network is important for memory and emotional processes. The limbic thalamus consists of the anterior thalamic nuclei, the lateral dorsal nucleus, and the medial-dorsal nucleus (Aggleton et al, 2016; Jones Mateen, Lucchinetti, Jack, & Welker, 2011). The limbic thalamus acts as a hub for spatial orientation and episodic memory. Patients with AD show episodic memory and spatial orientation impairment due to damage to the limbic thalamus and the Papez circuit. Aggleton et al. (2016) showed that a loss of episodic memory reflects greater neurodegeneration, which includes the Papez circuit and the limbic thalamus. The anterior thalamic nuclei are vital in this system.

One of the major sites of atrophy in AD is posterior cingulate region (Hornberger et al., 2012; Zhou et al., 2013). Importantly, Nestor, Fryer, and Hodges (2006) found that the episodic memory impairment in AD is linked with dysfunction in diencephalon and the posterior cingulate network. According to Aggleton and Brown (1999), the Papez circuit involves the hippocampus, fornix, mammillary bodies, anterior thalamic nuclei, and posterior cingulate region and is important for episodic memory. Reduced hippocampal functional connectivity was found in the thalamus in AD as the fornix undergoes atrophy, which leads to episodic memory impairment. Hornberger et al. (2012) mentioned that all areas of the Papez circuit except anterior cingulate region undergo degeneration in AD patients.

In addition, the anterior thalamus in Papez circuit also sends fibers to the hippocampus via the fornix. All these three areas—hippocampus, anterior thalamus, and fornix—undergo atrophy in AD. This can be viewed as the disruption of the Papez circuit as both the hippocampus and the posterior cingulate region are connected to the anterior thalamic nuclei, are a part of the Papez circuit, and undergo atrophy as a result of AD.

Retrosplenial cortex

The retrosplenial cortex includes the posterior cingulate region, which is vulnerable to atrophy and volume change in the early stages of AD (Aggleton et al., 2016). The posterior cingulate region becomes active when an individual performs tasks that require self-generated thoughts such as thinking of the future, episodic memory, and spatial navigation (Jones et al., 2011). The anterior and dorsomedial nuclei of the thalamus have strong connections with the posterior cingulate region. In AD, the presence of a lesion in the left anterior thalamus damages the posterior cingulate region portion of the default mode network (Jones et al., 2011).

The anterior thalamic nuclei and retrosplenial cortex are vital for episodic memory and are connected by dense connections (Aggleton & Brown, 1999). They also function interdependently for spatial learning and episodic memory and attention (Aggleton et al., 2016). Robertson and Kaitz (1981) found that the anterodorsal nucleus has connections with the retrosplenial limbic area and the anteroventral and lateral dorsal nuclei have connections with the cingulate area (Zarei et al., 2010). The presence of anterior thalamic lesions results in chronic dysfunction in the retrosplenial cortex (Torso et al., 2015). This dysfunction leads to disruption in gene transcription, which decreases the immediate early gene expression as in the hippocampus. It also leads to reduced metabolic activity and loss of some forms of neuronal plasticity (Aggleton et al., 2016).

Patients with AD found it difficult to change spatial orientation, which was linked to retrosplenial/posterior cingulate atrophy (Aggleton et al., 2016). This corroborates with the finding that determination of spatial orientation change requires hippocampus, posterior, cingulate cortex, anterior thalamus working in unison with retrosplenial cortex being the core of this interaction.

Thalamocortical network

The thalamocortical network consists of the thalamic nuclei and the cerebral cortex reciprocally connected to each other (Abuhassan, Coyle & Maguire, 2014). This connection is regulated by the reticular nuclei of the thalamus (Jagirdar & Chin, 2019). Reticular neurons are a thin sheet of inhibitory neurons that surround the anterior and lateral parts of thalamus and are very important for oscillations of thalamic neurons (Jagirdar & Chin, 2019). Disruption to the inhibitory function of these neurons impacts their oscillations (Abuhassan et al., 2014). The thalamocortical oscillations play a pivotal role in regulating cognitive processes and physical activity. According to Abuhassan et al. (2014), the oscillatory activity of these neurons in patients with AD was found to be abnormal. They also suggested that cognitive impairment in AD arises due to impaired connectivity in the thalamocortical network mainly between the medial thalamus and the cortical regions, the temporal, frontal, and occipital lobes (Abuhassan et al., 2014).

The thalamocortical network is also important for attentional processes and switching between sleep and wakefulness states as the sensory information has to pass through the thalamus before being relayed to the cortex (Jagirdar & Chin, 2019). In patients with AD, when the reticulate nuclei undergo loss in volume, their activity severely decreases,which impairs the communication between the thalamus and the cortex. The reticulate nuclei were also found to have great levels of dystrophic neurites that represent injured axon terminals. Therefore disruptions in the thalamocortical network due to AD leads to attentional deficits. Furthermore, this dysregulation to the corticothalamic network creates a discharge in spike wave, which leads to nonconvulsive seizures that is evident in patients with AD (Jagirdar & Chin, 2019). In addition, Abuhassan et al. (2014) found that thalamic activity severely decreases after partial corticothalamic denervation. This has led to the speculation that the observed reduction in the thalamic volume might be a result of corticothalamic denervation. Moreover, partial corticothalamic denervation also leads to decrease in thalamic oscillatory activity.

Prefrontal-striatal loops and the medial prefrontal cortex

The prefrontal-striatal network is a feed-forward loop that originates from the prefrontal cortex to the striatum (input structure of the basal ganglia), which then projects to the thalamus and back to the prefrontal region (Jeong et al., 2018). This network is essential for goal-directed behavior and executive function. A damage to this network leads to apathy, reduced empathy as a result of cognitive and emotional processing impairments (Levy & Dubois, 2005; Theleritis, Politis, Siarkos, & Lyketsos, 2013). Bertoux et al. (2015) found that in AD, there was significant posterior dorso-lateral prefrontal degeneration, but the prefrontal-striatal atrophy was relatively minimal. Degeneration of the thalamus due to AD affects the connectivity of the prefrontal-striatal network possibly leading to cognitive impairments and apathy as observed in patients with AD. In addition, the striatum has a critical role in cognition and behavior and also is important in facilitating voluntary movement (Bertoux et al., 2015). The striatum includes the bilateral caudate nucleus and putamen. Patients with AD showed striatal atrophy (Bertoux et al., 2015; Yi, Bertoux, Mioshi, Hodges, & Hornberger, 2013). Research conducted by Jeong et al. (2018) found that AD patients with apathy showed lower cerebral perfusion in orbitofrontal cortex, the thalamus, putamen, nucleus accumbens, and insula when compared to patients without apathy. In addition, it was found that a reduction in regional cerebral blood flow directly correlated with serious apathy.

The orbitofrontal cortex has dense connections with limbic/paralimbic structures and receives signals from brain regions including primary sensory organs, hypothalamus, and the limbic system (Kobayashi, 2009; Naish & Balodis, 2019). The orbitofrontal cortex is an integral part of the prefrontal-striatal network (Jeong et al., 2018). It is involved in making everyday decisions and plays an important role in reward processes (Balasubramani, Chakravarthy, Balaraman, & Moustafa, 2014, 2015; Mandali, Rengaswamy, Chakravarthy, & Moustafa, 2015; Moustafa & Gluck, 2011a, 2011b). In addition, it also combines sensory, affective, and motivational information to evaluate action and reward (Wallis, 2007). Damage to the orbitofrontal cortex leads to impairment in day-to-day decision-making and inhibits response (Theleritis et al., 2013). This matches characteristic of apathy. Hypoperfusion, hypometabolism, and atrophy of this orbitofrontal cortex have been reported to be related to apathy in AD patients (Theleritis et al., 2013).

The medial prefrontal cortex (mPFC) combines information from various cortical and subcortical areas. It is crucial for different functions including cognitive processes, emotion regulation, motivation, and working memory (Xu et al., 2019). The mPFC is classified into four subregions along the dorsal and the ventral axis: medial precentral area, anterior cingulate cortex, prelimbic cortex, and infralimbic cortex (Heidbreder & Groenewegen, 2003). The mPFC primarily consists of pyramidal neurons. Pyramidal neurons in the PL and the IL receive input from regions including the midline thalamus, basolateral amygdala, ventral hippocampus, and contralateral mPFC (Xu et al., 2019). The thalamus projects to the first layer of both dorsal and ventral PFC.

It has been shown that the symptoms of AD are related to pathological/functional changes in the mPFC. The default mode network has structural interconnections with mPFC, posterior cingulate cortex and the anterior and dorsomedial thalamic nuclei (Jones et al., 2011; Xu et al., 2019). The functional connectivity of the default mode network with the thalamic nuclei and the mPFC and its activity both decline in patients with mild AD (Xu et al., 2019). AD leads to accumulation of extracellular Aβ plaques in the mPFC. The presence of the plaques leads to poor behavioral performance and induced working memory deficits in rats (Xu et al., 2019). The decrease in connection between mediodorsal thalamus and the mPFC due to AD also leads to impairment in the working memory. In addition, the working memory and cognitive deficit seen in patients with AD is also due to the decrease in the spine density of the mPFC pyramidal neurons and in the hippocampus. The decline of spine density leads to decline of working memory as the dendritic spine density increases after memory consolidation (Xu et al., 2019).

Discussion

AD is a common neurogenerative disorder that results in the formation of neurofibrillary tangles and amyloid plaques in the brain (De Jong et al., 2008). The thalamus is one of the earliest brain regions to be affected by amyloid deposition in AD (Ryan et al., 2013). Our review shows that AD impacts both the thalamus itself (e.g., decrease in volume and cell loss), as well as thalamus's connections to other brain regions, including hippocampus, Papez circuit, the retrosplenial cortex, and other cortical areas. In AD, the thalamic nuclei are severely affected as they undergo loss in volume and degeneration. According to Zidan et al. (2019), thalamic volume loss can be attributed to the amyloid accumulation and axon degeneration in patients with AD. Xuereb et al. (1991) concluded that all the thalamic nuclei except the lateral posterior and central lateral nuclei show reductions in volume in AD with the reticular nucleus, dorsal nucleus, and the ventral thalamic nuclei showing significant reduction in volume (Low et al., 2019). In addition, it was found that the anterodorsal nuclei in the anterior nuclear group undergoes the greatest nerve cell loss along with the centromedial nucleus and the pulvinar nuclei. The centromedial nucleus, however, lacks significant tangle formation, which is evident in the anterior nuclear group (Aggleton et al., 2016; Xuereb et al., 1991; Zarei et al., 2010).

However, factors underlying nerve cell loss and its consequences is still unknown and much research needs to be conducted to better understand these factors (Xuereb et al., 1991; Zarei et al., 2010). Moreover, nerve cells present in the centromedial nuclei of the thalamus were found to get reduced to dark and shrunken structures. This is linked to the increased number of astrocytes, which indicates antemortem neural damage. However, this finding cannot be solely due to the development of AD, as elderly healthy individuals also had equally severe nerve cell loss in the same thalamic region (Xuereb et al., 1991). In addition, Aggleton et al. (2016) concluded that the anteroventral thalamic nuclei also experiences nerve cell loss contrary to Xuereb et al.'s (1991) findings that the anteroventral and anteromedial nuclei do not undergo nerve cell loss. Zarei et al. (2010) also found that the intralaminar nuclei were found to be degenerated in early stages of AD. It is involved in memory process and attention and forms cortico-thalamo-cortical pathways (De Medeiros Silva et al., 2014) In addition, the thalamus is connected to other regions of the brain such as the hippocampus, the retrosplenial cortex and this connection is impacted in AD. The hippocampus has direct connections with the anterior thalamus via the fornix and to the pulvinar through the temporo-pulvinar tract. This connection is important for retrieval of episodic memory and also involves the Papez circuit (Aggleton et al., 2016; Zarei et al., 2010). However, cell loss, neurofibrillary tangle formation, and atrophy is evident in the fornix during AD. This leads to disconnection from the hippocampal input leading to disruption in episodic memory. However, fornix shows cell loss and neurofibrillary tangle formation and atrophies during AD leading to disconnection from the hippocampal inputs. This leads to disruption in episodic memory (Aggleton et al., 2016).

Furthermore, the retrosplenial cortex that includes the posterior cingulate region is vital for episodic memory (Aggleton & Brown, 1999; Aggleton et al., 2016). This posterior cingulate region is subjected to atrophy and volume change in early stage of AD (Aggleton et al., 2016). In addition, the retrosplenial cortex has dense connections with the anterior thalamic nuclei (Aggleton & Brown, 1999). AD results in the presence of anterior thalamic lesions, leading to chronic dysfunction in the retrosplenial cortex, which further disrupts episodic memory. In addition, determining the change in spatial orientation requires hippocampus, cingulate cortex, and anterior thalamus working together. As their connections are damaged in AD, patients struggle with change in spatial orientation (Aggleton et al., 2016).

The limbic thalamus is essential for spatial orientation and episodic memory is connected to the Papez circuit. The Papez circuit involves the hippocampus and posterior cingulate region and also plays an important role in maintaining episodic memory (Aggleton et al, 2016; Jones et al., 2011). Aggleton and Brown (1999) and Hornberger et al. (2012) concluded that reduced hippocampal connectivity was found in the thalamus resulting in interruption in episodic memory. This shows the indirect impact AD has on the Papez circuit and its connections. It was also found that all areas of Papez circuit except anterior cingulate region undergo degeneration in AD (Hornberger et al., 2012). Lastly, the thalamocortical network connects the thalamic nuclei and the cortex and this network is regulated by the reticular nuclei of the thalamus (Abuhassan et al., 2014; Jagirdar & Chin, 2019). Reticular nuclei undergo loss in volume in people with AD. This change in volume of the nuclei results in disruption in the communication between the thalamus and the cortex (Jagirdar & Chin, 2019). This results in impaired connectivity between the thalamocortical network and the cortex leading to cognitive impairment. In addition, these reticular nuclei are vital for maintaining thalamic oscillations, which were found to be abnormal in patients with AD (Jagirdar & Chin, 2019). The thalamic oscillations play a crucial role in regulating cognitive and functional activity (Abuhassan et al., 2014). In addition, the thalamocortical network is also important for generating attention. Disruption in the thalamocortical network therefore leads to attention deficits in AD (Jagirdar & Chin, 2019).

Moreover, apathy has been recognized as a symptom of AD. The prefrontal-striatal network is a feed-forward loop that projects to the thalamus and is involved in executive function. Thalamic degeneration due to AD disrupts the connection with the prefrontal-striatal network resulting in cognitive impairment and apathy (Levy & Dubois, 2005; Theleritis et al., 2013). The orbitofrontal cortex is a crucial part of the prefrontal-striatal network and is important in making day-to-day decisions (Jeong et al., 2018; Wallis, 2007). Therefore damage to the orbitofrontal cortex results in difficulty in response inhibition and in decision-making (Theleritis et al., 2013).

The mPFC integrates information from various input regions and passes the updated information to the output regions. It plays an important role in many brain functions including cognitive processes, emotion regulation, and working memory (Xu et al., 2019). The default mode network has strong structural connection with the mPFC, posterior cingulate cortex, and the anterior and dorsomedial thalamic nuclei. Decline in this connectivity and formation of amyloid plaques in the mPFC due to AD leads to deterioration in behavioral performance and working memory (Xu et al., 2019). This affect in working memory can also be credited to decline in dendritic spine density of the mPFC pyramidal neurons due to AD (Xu et al., 2019). Thus AD greatly disrupts the thalamus by causing nerve cell loss, loss in volume, and affects its associated networks. However, more research is yet to be conducted to better understand the effect of AD on the thalamus.

Future work

The effect of AD on the thalamus is mostly overlooked and has not been extensively studied. It is still assumed that thalamic dysfunctions are a secondary response to disruption in medial temporal lobe and the result of cognitive loss due to thalamic dysfunction is underestimated (Aggleton et al., 2016). In addition, the importance of projections from the limbic thalamus to the hippocampus formation is also overlooked and need further investigation. Presently, magnetic resonance imaging of the nuclei of the thalamus have been performed to better understand the volume, neuron loss, and degeneration in the thalamus. Future studies that research on cellular, molecular, and circuitry aspects of corticothalamic function in AD patients' models can provide critical insight (Jagirdar & Chin, 2019). In addition, more investigation needs to be conducted to understand amyloid pathology across the Papez network. To gain a wider knowledge and a better understanding of the effect of thalamus in AD, much research needs to be conducted. A quantitative study of the thalamus in people of all ages, young, middle aged, and elderly needs to be conducted, which can then be contrasted with patients diagnosed with AD (Xuereb et al., 1991).

Conclusion

To conclude, the thalamus acts as relay for incoming sensory information and is affected by AD with the formation of neurofibrillary tangles and amyloid plaques along with synaptic loss in the thalamus. In addition, it also leads to decrease in the volume of thalamus mainly in the reticular and dorsal nucleus, which correlates to impaired cognitive performance. It also leads to nerve cell loss and degeneration mainly in the anterodorsal thalamic nuclei.

Furthermore, thalamus is strongly linked to other parts of the brain such as hippocampus, retrosplenial cortex, and cerebral cortex and has major roles in Papez circuit, the Default Mode Network, hippocampus, the thalamocortical network, and the prefrontal-striatal network. All these delicate processes are disrupted as the thalamus and other parts of the brain involved in different processes suffer atrophy because of AD.

The study of thalamic atrophy due to AD has not received much attention and much more research needs to be conducted if we are to fully understand in depth and gain insight regarding effect of thalamus on AD.

References

Abuhassan, K., Coyle, D., & Maguire, L. (2014). Compensating for thalamocortical synaptic loss in Alzheimer''s disease. *Frontiers in Computational Neuroscience, 8*, 65.

Aggleton, J., & Brown, M. (1999). Episodic memory, amnesia, and the hippocampal—anterior thalamic axis. *Behavioral and Brain Sciences, 22*(3), 425—444.

Aggleton, J., Pralus, A., Nelson, A., & Hornberger, M. (2016). Thalamic pathology and memory loss in early Alzheimer's disease: Moving the focus from the medial temporal lobe to Papez circuit. *Brain, 139*(7), 1877—1890.

Aisen, P., Cummings, J., Jack, C., Morris, J., Sperling, R., Frölich, L., et al. (2017). On the path to 2025: Understanding the Alzheimer's disease continuum. *Alzheimer's Research & Therapy, 9*(1).

Balasubramani, P. P., Chakravarthy, S., Balaraman, R., & Moustafa, A. A. (2014). An extended reinforcement learning model of basal ganglia to understand the contribution of serotonin and dopamine in risk-based decision making, reward prediction, and punishment learning. *Frontiers in Computational Neuroscience, 8*, 47.

Balasubramani, P. P., Chakravarthy, S., Balaraman, R., & Moustafa, A. A. (2015). A network model of basal ganglia for understanding the roles of dopamine and serotonin in reward-punishment-risk based decision making. *Frontiers in Computational Neuroscience, 9*, 76.

Bertoux, M., O'callaghan, C., Flanagan, E., Hodges, J., & Hornberger, M. (2015). Fronto-striatal atrophy in behavioral variant frontotemporal dementia and Alzheimer's disease. *Frontiers in Neurology, 6*, 147.

Blumenfeld, H., & Gummadavelli, A. 2018, Thalamus, Encyclopædia Britannica.

Cassel, J., Pereira De Vasconcelos, A., Loureiro, M., Cholvin, T., Dalrymple-Alford, J., & Vertes, R. (2013). The reuniens and rhomboid nuclei: Neuroanatomy, electrophysiological characteristics and behavioral implications. *Progress in Neurobiology, 111*, 34—52.

Cutsuridis, V., & A.A. Moustafa 2014. Computational modes of Alzheimer's disease, Scholarpedia, viewed April 8, 2020.

De Jong, L. W., Van Der Hiele, K., Veer, I. M., Houwing, J. J., Westendorp, R. G., Bollen, E. L., et al. (2008). Strongly reduced volumes of putamen and thalamus in Alzheimer's disease: An MRI study. *Brain: A Journal of Neurology, 131*(12), 3277—3285.

De Medeiros Silva, A., De Santana, M., De Góis Morais, P., De Sousa, T., Januário Engelberth, R., De Souza Lucena, E., et al. (2014). Serotonergic fibers distribution in the midline and intralaminar thalamic nuclei in the rock cavy (Kerodon rupestris). *Brain Research, 1586*, 99—108.

Fama, R., & Edith, V. S. (2015). Thalamic structures and associated cognitive functions: Relations with age and aging. *Neuroscience and Biobehavioral Reviews, 54*, 29—37.

Feng, F., Zhou, B., Wang, L., Yao, H., Guo, Y., An, N., et al. (2019). The correlation of functional connectivity and structural connectivity between hippocampus and thalamus in Alzheimer's disease and amnestic mild cognitive impairment. *Zhonghua Nei Ke Za Zhi [Chinese Journal of Internal Medicine], 58*(9), 662—667.

Gazzaniga, M., Ivry, R., & Mangun, G. (1998). *Cognitive neuroscience* (4th ed., pp. 45—46). New York: W.W. Norton and Company.

Hampel, H., Mesulam, M. M., Cuello, A. C., Farlow, M. R., Giacobini, E., Grossberg, G. T., et al. (2018). The cholinergic system in the pathophysiology and treatment of Alzheimer's disease. *Brain*, *141*, 1917–1933.

Heidbreder, C., & Groenewegen, H. (2003). The medial prefrontal cortex in the rat: Evidence for a dorso-ventral distinction based upon functional and anatomical characteristics. *Neuroscience & Biobehavioral Reviews*, *27*(6), 555–579.

Hodson, R. (2018). Alzheimer's disease. *Nature*, *559*(7715).

Hornberger, M., Wong, S., Tan, R., Irish, M., Piguet, O., Kril, J., et al. (2012). In vivo and post-mortem memory circuit integrity in frontotemporal dementia and Alzheimer's disease. *Brain*, *135*(10), 3015–3025.

Huerta-Ocampo, I., Hacioglu-Bay, H., Dautan, D., & Mena-Segovia, J. (2020). Distribution of midbrain cholinergic axons in the thalamus. *eNeuro*, *7*.

Iglesias, J., Insausti, R., Usabiaga, G., Bocchetta, M., Leemput, K., Greve, D., et al. (2018). A probabilistic atlas of the human thalamic nuclei combining ex vivo MRI and histology. *Neuroimage*, *183*, 314–326.

Jagirdar, R., & Chin, J. (2019). Corticothalamic network dysfunction and Alzheimer's disease. *Brain Research*, *1702*, 38–45.

Jenkins, T., Amin, E., Brown, M., & Aggleton, J. (2006). Changes in immediate early gene expression in the rat brain after unilateral lesions of the hippocampus. *Neuroscience*, *137*(3), 747–759.

Jeong, H., Kang, I., Im, J., Park, J., Na, S., Heo, Y., et al. (2018). Brain perfusion correlates of apathy in Alzheimer's disease. *Dementia and Neurocognitive Disorders*, *17*(2), 50.

Jones, D., Mateen, F., Lucchinetti, C., Jack, C., & Welker, K. (2011). Default mode network disruption secondary to a lesion in the anterior thalamus. *Archives of Neurology*, *68*(2).

Kanel, P., Muller, M., Van Der Zee, S., Sanchez-Catasus, C. A., Koeppe, R. A., Frey, K. A., & Bohnen, N. I. (2020). Topography of cholinergic changes in dementia with Lewy bodies and key neural network hubs. *The Journal of Neuropsychiatry and Clinical Neurosciences*, *32*(4), 370–375, appineuropsych19070165.

Kobayashi, S. (2009). *Reward neurophysiology and primate cerebral cortex. Encyclopedia of Neuroscience* (pp. 325–333). Academic Press.

Kotagal, V., Muller, M. L., Kaufer, D. I., Koeppe, R. A., & Bohnen, N. I. (2012). Thalamic cholinergic innervation is spared in Alzheimer disease compared to parkinsonian disorders. *Neuroscience Letters*, *514*(2), 169–172. Available from https://doi.org/10.1016/j.neulet.2012.02.083.

Koyama, M. S., Molfese, P. J., Milham, M. P., Mencl, W. E., & Pugh, K. R. (2020). Thalamus is a common locus of reading, arithmetic, and IQ: Analysis of local intrinsic functional properties. *Brain and Language*, *209*, 104835.

Lane, C., Hardy, J., & Scott, J. (2018). Alzheimer's disease. *European Journal of Neurology*, *25*, 59–70.

Levy, R., & Dubois, B. (2005). Apathy and the functional anatomy of the prefrontal cortex–basal ganglia circuits. *Cerebral Cortex*, *16*(7), 916–928.

Li, X., Xia, J., Ma, C., Chen, K., Xu, K., Zhang, J., et al. (2020). Accelerating structural degeneration in temporal regions and their effects on cognition in aging of MCI patients. *Cerebral Cortex*, *30*, 326–338.

Li, Y., Lopez-Huerta, V. G., Adiconis, X., Levandowski, K., Choi, S., Simmons, S. K., et al. (2020). Distinct subnetworks of the thalamic reticular nucleus. *Nature*, *583*, 819–824.

Low, A., Mak, E., Malpetti, M., Chouliaras, L., Nicastro, N., Su, L., et al. (2019). Asymmetrical atrophy of thalamic subnuclei in Alzheimer's disease and amyloid-positive mild cognitive impairment is associated with key clinical features. *Alzheimer's & Dementia: Diagnosis, Assessment & Disease Monitoring*, *11*(1), 690–699.

Mandali, A., Rengaswamy, M., Chakravarthy, S., & Moustafa, A. A. (2015). A spiking basal ganglia model of synchrony, exploration and decision making. *Frontiers in Neuroscience*, *9*, 191.
Mega, M. S. (2000). The cholinergic deficit in Alzheimer's disease: Impact on cognition, behaviour and function. *The International Journal of Neuropsychopharmacology/Official Scientific Journal of the Collegium Internationale Neuropsychopharmacologicum (CINP)*, *3*(7), 3–12. Available from https://doi.org/10.1017/S1461145700001942.
Minter, M., Taylor, J., & Crack, P. (2015). The contribution of neuroinflammation to amyloid toxicity in Alzheimer"s disease. *Journal of Neurochemistry*, *136*(3), 457–474.
Moustafa, A. A., & Gluck, M. A. (2011a). Computational cognitive models of prefrontal-striatal-hippocampal interactions in Parkinson's disease and schizophrenia. *Neural Networks*, *24*(6), 575–591.
Moustafa, A. A., & Gluck, M. A. (2011b). A neurocomputational model of dopamine and prefrontal-striatal interactions during multi-cue category learning by Parkinson's patients. *Journal of Cognitive Neuroscience*, *23*(1), 51–67.
Moustafa, A. A., Mcmullan, R. D., Rostron, B., Hewedi, D. H., & Haladjian, H. H. (2017). The thalamus as a relay station and gatekeeper: Relevance to brain disorders. *Reviews in the Neurosciences*, *28*(2), 203–218.
Murray Sherman S. 2006. Thalamus, *Scholarpedia*, viewed April 6, 2020, http://www.scholarpedia.org/article/Thalamus.
Naish, K., & Balodis, I. (2019). Reward processing in food addiction and overeating. In P. Cottone, C. Moore, V. Sabino, & G. Koob (Eds.), *Compulsive eating behavior and food addiction* (pp. 217–249). Elsevier.
Nestor, P., Fryer, T., & Hodges, J. (2006). Declarative memory impairments in Alzheimer's disease and semantic dementia. *Neuroimage*, *30*(3), 1010–1020.
Noftz, W. A., Beebe, N. L., Mellott, J. G., & Schofield, B. R. (2020). Cholinergic projections from the pedunculopontine tegmental nucleus contact excitatory and inhibitory neurons in the inferior colliculus. *Frontiers in Neural Circuits*, *14*, 43.
Pelzer, E. A., Pauls, K. A. M., Braun, N., Tittgemeyer, M., & Timmermann, L. (2020). Probabilistic tractography in the ventrolateral thalamic nucleus: Cerebellar and pallidal connections. *Brain Structure and Function*, *225*, 1685–1689.
Pita-Almenar, J. D., Yu, D., Lu, H. C., & Beierlein, M. (2014). Mechanisms underlying desynchronization of cholinergic-evoked thalamic network activity. *The Journal of Neuroscience*, *34*, 14463–14474.
Robertson, R., & Kaitz, S. (1981). Thalamic connections with limbic cortex. I. Thalamocortical projections. *The Journal of Comparative Neurology*, *195*(3), 501–525.
Ryan, N., Keihaninejad, S., Shakespeare, T., Lehmann, M., Crutch, S., Malone, I., et al. (2013). Magnetic resonance imaging evidence for presymptomatic change in thalamus and caudate in familial Alzheimer's disease. *Brain*, *136*(5), 1399–1414.
Sheridan, N., & Tadi, P. (2020). *Neuroanatomy, thalamic nuclei*. Treasure Island, FL: StatPearls.
Štillová, K., Jurák, P., Chládek, J., Chrastina, J., Halámek, J., Bočková, M., et al. (2015). The role of anterior nuclei of the thalamus: A subcortical gate in memory processing: An intracerebral recording study. *PLoS One*, *10*(11), e0140778.
Taber, K., Wen, C., Khan, A., & Hurley, R. (2015). The limbic thalamus. *Journal of Neuropsychiatry*, *16*(2), 127–132.
Theleritis, C., Politis, A., Siarkos, K., & Lyketsos, C. (2013). A review of neuroimaging findings of apathy in Alzheimer's disease. *International Psychogeriatrics*, *26*(2), 195–207.
Torso, M., Serra, L., Giulietti, G., Spanò, B., Tuzzi, E., Koch, G., et al. (2015). Strategic lesions in the anterior thalamic radiation and apathy in early Alzheimer's disease. *PLoS One*, *10*(5), e0124998.
Vertes, R., Linley, S., & Hoover, W. (2015). Limbic circuitry of the midline thalamus. *Neuroscience & Biobehavioral Reviews*, *54*, 89–107.

Wallis, J. (2007). Orbitofrontal cortex and its contribution to decision-making. *Annual Review of Neuroscience*, *30*(1), 31–56.

Xu, P., Chen, A., Li, Y., Xing, X., & Lu, H. (2019). Medial prefrontal cortex in neurological diseases. *Physiological Genomics*, *51*(9), 432–442.

Xuereb, J. H., Perry, R. H., Candy, J. M., Perry, E. K., Marshall, E., & Bonham, J. R. (1991). Nerve cell loss in the thalamus in Alzheimer's and Parkinson's disease. *Brain*, *114*(Pt 3), 1363–1379.

Yi, D., Bertoux, M., Mioshi, E., Hodges, J., & Hornberger, M. (2013). Fronto-striatal atrophy correlates of neuropsychiatric dysfunction in frontotemporal dementia (FTD) and Alzheimer's disease (AD). *Dementia and Neuropsychologia*, 7(1), 75–82.

Zarei, M., Patenaude, B., Damoiseaux, J., Morgese, C., Smith, S., Matthews, P., et al. (2010). Combining shape and connectivity analysis: An MRI study of thalamic degeneration in Alzheimer's disease. *Neuroimage*, *49*(1), 1–8.

Zhou, B., Liu, Y., Zhang, Z., An, N., Yao, H., Wang, P., et al. (2013). Impaired functional connectivity of the thalamus in Alzheimer's disease and mild cognitive impairment: A resting-state fMRI study. *Current Alzheimer Research*, *10*(7), 754–766.

Zidan, M., Boban, J., Bjelan, M., Todorović, A., Stankov Vujanić, T., Semnic, M., et al. (2019). Thalamic volume loss as an early sign of amnestic mild cognitive impairment. *Journal of Clinical Neuroscience*, *68*, 168–173.

CHAPTER 7

Using big data methods to understand Alzheimer's disease

Samuel L. Warren[1], Ahmed A. Moustafa[2,3,4] and Hany Alashwal[5]
[1]Psychological Science, School of Psychology, Western Sydney University, Sydney, NSW, Australia
[2]School of Psychology, Western Sydney University, Sydney, NSW, Australia
[3]MARCS Institute for Brain and Behaviour, Western Sydney University, Sydney, NSW, Australia
[4]Department of Human Anatomy and Physiology, The Faculty of Health Sciences, University of Johannesburg, Johannesburg, South Africa
[5]College of Information Technology, United Arab Emirates University, Al-Ain, United Arab Emirates

Introduction

Alzheimer's disease is the most common form of dementia that affects 50 million people worldwide (World Health Organization, 2017). In the next three decades, dementia diagnoses are predicted to drastically increase to 132 million cases globally (Alzheimer's Disease International, 2019), causing an Alzheimer's disease epidemic. The rapid increase in dementia cases is problematic, as Alzheimer's disease currently has no cure and always results in death (Alzheimer's Association, 2018). Consequentially, Alzheimer's disease is the fifth leading cause of death worldwide (Alzheimer's Association, 2018; World Health Organization, 2018). It is safe to say that Alzheimer's disease is one of the greatest medical threats of our time. Accordingly, there is a critical need for Alzheimer's disease research to overcome what is expected to be a global epidemic.

Throughout this article, we have explored how big data can analyze and overcome complex problems. In this chapter, we explore how big data is applied to Alzheimer's disease research. We aim to provide a general understanding of Alzheimer's disease and outline common methods of big data analysis in dementia research. We also explore big data collection and describe the advantages of databases over standard psychological research. At the end of the chapter, we discuss the *Alzheimer's Disease Neuroimaging Initiative* (ADNI) database as a specific example of big data analysis in Alzheimer's disease research. Accordingly, we aim to highlight the unique advantages of big data analysis and discuss how researchers are using big data to combat Alzheimer's disease.

Alzheimer's Disease
DOI: https://doi.org/10.1016/B978-0-12-821334-6.00004-1

This chapter consists of four sections that broadly discuss Alzheimer's disease, big data methods, predictive modeling, and the ADNI. Specifically, each section discusses the following topics: (1) section "What is Alzheimer's disease?" provides a brief introduction to Alzheimer's disease and explores the causes, treatments, diagnosis, and study of the disorder; (2) section "Big data methods" highlights the advantages of big data methods and outlines the role of databases in Alzheimer's disease research; (3) section "Big data analysis" discusses the importance of predictive models (statistical and computational models) in the study of Alzheimer's disease; and (4) section "Example: the Alzheimer's Disease Neuroimaging Initiative" explores the methods and impact of the ADNI database. In turn, we aim to provide a general understanding of how big data is applied to Alzheimer's disease research.

What is Alzheimer's disease?

Alzheimer's disease involves the loss of mental and physical function due to the gradual death of the brain (Dubois et al., 2010). Individuals with Alzheimer's disease will commonly experience symptoms of memory loss, communicative difficulty, confusion, and loss of movement (American Psychiatric Association, 2013). Memory loss is the defining characteristic of Alzheimer's disease as the brain starts decaying in regions associated with memory function (Bastin & Salmon, 2014; Perry, Watson, & Hodges, 2000); however, many other cognitive and physical abilities (e.g., language and motor skills) are also affected. As Alzheimer's disease progresses the symptoms of the disease become more severe. For example, individuals with late-stage Alzheimer's disease are often unable to perform tasks such as walking or speaking. Moreover, as Alzheimer's disease progresses the chances of medical complications occurring (e.g., pneumonia) also increases (Doraiswamy, Leon, Cummings, Marin, & Neumann, 2002; Wada et al., 2001). Accordingly, individuals with Alzheimer's disease will experience a gradual loss of function to the point of death.

What causes Alzheimer's disease?

The cause of Alzheimer's disease is currently unknown. However, current theories, such as the amyloid cascade hypothesis, provide great insight into the progression of Alzheimer's disease. The amyloid cascade hypothesis states that Alzheimer's disease development is related to the aggregation and formulation of toxic matter known as senile plaques and neurofibrillary tangles (Bloom, 2014). It is theorized that high concentrations of the peptide amyloid-beta (Aβ) results in the formulation of senile plaques (Sharma & Singh, 2016). Senile plaques are inherently toxic and, when accumulated, cause neuron (brain cell) death (Yankner, Duffy, & Kirschner, 1990). The accumulation of senile plaques also alters the protein tau resulting in the formulation of neurofibrillary tangles (Iqbal, Liu, Gong, & Grundke-Iqbal, 2010). Neurofibrillary tangles compromise the structural integrity of neurons resulting in cell death and further toxicity (Jack et al., 2010). Accordingly, both Aβ and tau perpetuate a cycle of neurotoxicity and compromise neurons to the point of death.

The amyloid cascade hypothesis also states that neuron loss starts in the hippocampus and then progressively expands into the wider medial temporal lobe and cortical regions of the brain (El Haj et al., 2016). This cascading degeneration has been observed in the literature and is fundamental to the current understanding of Alzheimer's disease progression (Murphy & LeVine, 2010). Subsequently, the amyloid cascade hypothesis is the current theoretical basis for most Alzheimer's disease research. However, it is important to note that the amyloid cascade hypothesis is not an exhaustive theory of Alzheimer's disease. For example, studies have shown that healthy adults can have high levels of Aβ and not develop Alzheimer's disease (Rodrigue et al., 2012). Moreover, the cause of Aβ and tau aggregation is also unknown. Accordingly, additional theories of Alzheimer's disease development are also used in research (e.g., the cholinergic hypothesis) (Bartus, Dean, Beer, & Lippa, 1982; Craig, Hong, & McDonald, 2011); however, alternative theories are beyond the scope of this chapter.

Are there treatments for Alzheimer's disease?

While there are no cures for Alzheimer's disease, there are treatments available. Specifically, drugs and aged care are commonly used to counteract the symptoms of Alzheimer's disease (Feldman et al., 2003). The focus of aged care is to help individuals with daily tasks that are difficult due to Alzheimer's disease decline (Bradway et al., 2012). For example, carers will often help feed, clean, and medicate individuals with Alzheimer's disease. It is important to note that individuals require different levels of care depending on the severity of Alzheimer's disease. Through aged care assistance, individuals with Alzheimer's disease can remain in a comfortable environment longer (e.g., retirement village) and maintain some autonomy (Edvardsson, Fetherstonhaugh, & Nay, 2010). Accordingly, aged care is instrumental in extending and improving the quality of life of individuals living with Alzheimer's disease.

Drugs are also used to treat the symptoms of Alzheimer's disease. For example, cholinesterase inhibitors are a type of drug commonly used to treat the cognitive symptoms of Alzheimer's disease (e.g., memory loss) (Tan et al., 2014). Cholinesterase inhibitors work by regulating the neurochemical acetylcholine, which is instrumental in cognitive function and is also depleted in Alzheimer's disease (Cummings et al., 2016). By regulating acetylcholine, cholinesterase inhibitors help maintain cognitive function and slow the symptoms of Alzheimer's disease (Ma, Ji, Li, Yang, & Pan, 2018). However, the ability of cholinesterase inhibitors varies depending on the individual, the stage of the disease, and the time of diagnosis (Small & Bullock, 2011). There are also few alternatives to cholinesterase inhibitors when treating Alzheimer's disease (Yiannopoulou & Papageorgiou, 2013). Accordingly, there is a great need for improvements in Alzheimer's disease medications; however, drug treatments are currently limited by Alzheimer's disease diagnosis.

Alzheimer's disease is currently diagnosed through a clinical evaluation. A doctor will use multiple psychological and physiological tests (e.g., memory, language, concentration, balance, reflexes, and motor skills) to assess for symptoms of Alzheimer's disease (McKhann et al., 2011). However, an Alzheimer's disease diagnosis can only be fully confirmed after death via an autopsy. Moreover, Alzheimer's disease diagnoses are also limited because they rely on symptoms of the disease that overlap with other forms of dementia (e.g., vascular dementia) (Brown, 2015; Larner, 2019). Accordingly, current Alzheimer's disease diagnoses occur at a later stage of the disease, when brain damage is presumed to be irreversible and most medications are largely ineffective (Kalat, 2018, p. 206). Current treatments, such as cholinesterase inhibitors, can mostly treat the symptoms of Alzheimer's disease (e.g., memory and motor function loss) but cannot reverse or prevent the disease (Edvardsson, Winblad, & Sandman, 2008; Tan et al., 2014). Accordingly, current research seeks to diagnose Alzheimer's disease at an earlier stage to improve treatment strategies.

How do researchers study Alzheimer's disease?

Current psychological research uses predictive methods to understand the progression of Alzheimer's disease. In turn, these predictors of Alzheimer's disease inform diagnosis strategies and potential treatment areas. Researchers commonly predict disease development and progression by studying disorders that transition to Alzheimer's disease. Specifically, the

Alzheimer's Association and the National Institute on Aging recognize Alzheimer's disease as a continuum with three different stages known as *preclinical Alzheimer's disease*, *mild cognitive impairment*, and *Alzheimer's dementia* (Jack et al., 2011). Preclinical Alzheimer's disease is the initial stage of Alzheimer's disease development and is an asymptomatic state that can last decades before transitioning to mild cognitive impairment or Alzheimer's disease (Dubois et al., 2010). Due to the nature of preclinical Alzheimer's disease, mild cognitive impairment is preferred in Alzheimer's disease research as the disorder is easier to predict, recognize, and study.

Mild cognitive impairment is defined as the nondemented impairment of at least one cognitive ability, such as language or memory (Csukly et al., 2016). It is important to note that these impairments are not debilitating in mild cognitive impairment. Debilitating impairments indicate another disorder such as dementia or Parkinson's disease. Moreover, mild cognitive impairment is not exclusive to Alzheimer's disease development. However, mild cognitive impairment does share many characteristics with Alzheimer's disease and reliably converts to the disease (Albert et al., 2011). Specifically, individuals with mild cognitive impairment convert to Alzheimer's disease at a rate of 10%–15% while only 2%–4% of the general population progress to Alzheimer's disease (Roberts et al., 2008). Accordingly, by studying mild cognitive impairment, scientists can guarantee a larger and more reliable population of Alzheimer's disease converters for research.

Problems with standard Alzheimer's disease research

However, studying participants with Alzheimer's disease and mild cognitive impairment can be incredibly complicated. Accordingly, Alzheimer's disease and mild cognitive impairment research is incredibly complex and is often limited by three main factors. First, it is incredibly hard to organize and recruit vulnerable populations for research (e.g., individuals with dementia). For example, neurological participants, such as participants with Alzheimer's disease, are likely to drop out of studies as the symptoms of their disease worsen (Watson, Ryan, Silverberg, Cahan, & Bernard, 2014). Second, Alzheimer's disease detection methods often have a high monetary and administrative cost that restricts their use. For example, magnetic resonance imaging (MRI) technology can measure neuron loss with high accuracy, yet the device is incredibly expensive to use and hard to access (Stites, Milne, & Karlawish, 2018). Consequentially, progress in Alzheimer's disease research may be delayed due to the inaccessibility of innovative technology.

Finally, the resources required to study Alzheimer's disease progression and mild cognitive impairment conversion longitudinally (long term) are immense and beyond the scope of some laboratories. Most, if not all laboratories, cannot afford to devote years of work to the longitudinal collection of Alzheimer's disease data. Accordingly, longitudinal topics like Alzheimer's disease progression can lack sufficient research. However, these resource, temporal, and recruitment barriers to Alzheimer's disease research can often be overcome through big data.

Big data methods

Alzheimer's disease is an inherently complex disorder that occurs across the lifespan. For example, some studies have estimated that signs of Alzheimer's disease, such as neurofibrillary tangles, can be observed decades before the onset of the disorder (Braak & Del Tredici, 2011). Accordingly, significant resources and specialized methods are required to longitudinally collect Alzheimer's disease data. In Alzheimer's disease research, big data commonly uses databases to record and maintain complex longitudinal datasets. In turn, these databases enable researchers to analyze Alzheimer's disease development across the lifespan. In this section, we discuss the background and advantages of databases in Alzheimer's disease research.

What are databases?

Databases are information hubs where researchers can upload, combine, share, and access Alzheimer's disease data (Mueller et al., 2005). Databases are often run by universities and organizations that seek to facilitate the collaborative collection of data. These databases are ongoing and therefore build large longitudinal datasets over time. It is common for researchers to freely use and contribute to a database. Accordingly, databases are essentially large collaborative datasets that provide rich longitudinal data to researchers. Databases also provide researchers with unique opportunities to access large Alzheimer's disease samples and a multitude of variables that would usually be unobtainable to an individual laboratory. Accordingly, the complex data contained in a database enables researchers to longitudinally assess and understand Alzheimer's disease.

For example, the ADNI is the largest database used in Alzheimer's disease research. As an organization, ADNI seeks to understand, detect, and observe Alzheimer's disease progression using disease biomarkers (biological indicators of Alzheimer's disease) (Alzheimer's Disease Neuroimaging

Initiative, 2017a). Specifically, ADNI collects neuroimaging, biomarker, psychological, and demographic measures in a cohort of control (neurologically healthy), mild cognitive impairment, and Alzheimer's disease participants (Alzheimer's Disease Neuroimaging Initiative, 2017b). Most importantly for researchers, ADNI operates on a policy of information sharing, which enables any study with an institutional ethics approval to use ADNI data. Through ADNI, researchers can study Alzheimer's disease while overcoming the logistical, temporal, and economic limitations discussed above. Accordingly, databases such as ADNI play a fundamental role in enabling and accelerate Alzheimer's disease research.

The advantages of databases

Databases have unique advantages that can often overcome the resource, recruitment, and temporal limitations of standard Alzheimer's disease research. Specifically, collaboration and information sharing enable researchers to overcome resource limitations, recruitment barriers, and conduct longitudinal research.

Standard psychological studies are often limited by the high monetary and administrative costs of Alzheimer's disease research. However, big data research often overcomes these *resource limitations* through databases. Databases have the unique ability to collect massive amounts of data in a short amount of time through collaboration (Papale et al., 2012). Collaborative data collection involves the cooperation of multiple research groups in creating an overarching database. Through collaboration, the resulting database contains more data than any one study could gather (Pearlson, 2009). Accordingly, researchers can access more data and overcome resource limitations. Databases also enable researchers to study Alzheimer's disease using technology and resources that are not easily affordable or accessible (Olfson, Wall, & Blanco, 2017). For example, researchers may be ill-equipped or unable to perform invasive procedures such as lumbar punctures (used to collect Aβ and tau); however, researchers can access similar data through a database and thus continue their work. Accordingly, costly and time-consuming research has been popularized in the literature through collaborative research and databases.

Databases can also overcome *recruitment barriers* (e.g., recruiting vulnerable populations), which commonly limit Alzheimer's disease research. Specifically, databases can overcome recruitment barriers through information sharing (Toga, Bhatt, & Ashish, 2016). Information sharing allows for the free access and distribution of Alzheimer's disease data. It is common for databases to freely share their data with the wider scientific community to accelerate the study of Alzheimer's disease. Through information sharing, researchers can access large preexisting databases and, thus avoid participant recruitment barriers (Martone, Garcia-Castro, & VandenBos, 2018). In return, databases can ensure a thorough and organized investigation of Alzheimer's disease. While information sharing does not remove the need for data collection, it does reduce the need of an individual researcher to recruit participants. In part, individual researchers can collect less data because of the collaborative collection and the efficient reuse of data by databases. Accordingly, information sharing enables researchers to focus their resources on treating, diagnosing, and understanding Alzheimer's disease rather than data collection (Geerts et al., 2016).

Another problem with standard psychological research is the inability of studies to observe Alzheimer's disease *longitudinally*. With current data collection methods, it could genuinely take a lifetime to study Alzheimer's disease progression. Moreover, securing stable funding for longitudinal Alzheimer's disease research is highly difficult. In big data research, databases can efficiently collect longitudinal Alzheimer's disease data through collaboration (Toga et al., 2016; Vikström, Edvinsson, & Brändström, 2002). It is apparent that when initiatives and organizations carry the collaborative burden, individual researchers do not need to devote years of their life to data collection (Geerts et al., 2016). In turn, studies can research Alzheimer's disease longitudinally through databases without worrying about data collection (Cook, Andriole, Durning, Roberts, & Triola, 2010). Accordingly, databases actively promote complex longitudinal research through the accessibility, quantity, and quality of their data.

Common difficulties with databases

Clearly, databases have unique advantages that help forward Alzheimer's disease research. However, databases are also very complex and require constant organization. In Alzheimer's disease research, databases contain a large amount of information that can be easily mismanaged. For example, when data are collected by different groups, measurements and variables can be incompatible or conflict. Organizational breakdowns in data collection directly threaten the collaborative notion of databases and can limit big data analyses. For example, missing data limits the statistical power and methods available when performing a big data analysis (Nakagawa & Freckleton, 2008). Accordingly, it is important that databases unify their data collection strategies to ensure efficient research.

Standardization is the process of conforming all scientific measures to the same rules and processes. The standardization of measures enables uniformity in data collection and database formulation (Papale et al., 2012). Accordingly, databases with multiple contributors should define a set of measures and rules that researchers must follow when collecting data. By following the same rules, studies can ensure that all measures are collected similarly and combine to make a cohesive dataset (Schmitt & Burchinal, 2011). Standardized measures and variables also ensure statistical power and decrease the chances of complications during the analysis. Moreover, the rules and methods of data collection outlined through standardization also enable future researchers to replicate and confirm studies' results. Accordingly, researchers who practice standardization will ensure efficiency in big data collection, decrease complications during big data analysis, and ensure transparency for future replication.

Big data analysis

Big data analysis is the investigation and interpretation of large datasets (Fan, Han, & Liu, 2014). In Alzheimer's disease research, big data commonly analyzes complex longitudinal datasets obtained from databases. For example, predictive models are often used to study complex longitudinal datasets obtained from the ADNI database (Moradi, Hallikainen, Hänninen, Tohka, & Alzheimer's Disease Neuroimaging Initiative, 2017). The use of predictive models is not unique to big data analysis and is widely used throughout research; however, big data does use unique statistical and computational techniques that specialize in analyzing vast amounts of data (Geerts et al., 2016). In this section, we discuss how statistical and computational models are used to study Alzheimer's disease. We also discuss how big data analysis can be used to help diagnose and treat Alzheimer's disease in clinical practice.

Statistical modeling

In big data analysis, statistical models use mathematical patterns to predict future outcomes (Eaton & Baltes, 2001). In Alzheimer's disease research, statistical models are commonly used to determine the importance of a variable in the development of the disease. For example, researchers will often observe disease biomarkers over time (e.g., Aβ accumulation) to predict Alzheimer's disease development (Forlenza et al., 2010). The predictive variables that make up a statistical model are usually determined by theory or previous Alzheimer's disease research. When conducting an analysis, the accuracy and efficacy of a statistical model are determined by testing the model against the observed outcomes in the dataset (Mayer & Butler, 1993). For example, if researchers are trying to predict Alzheimer's disease development, the statistical model will be compared to the observed disease development in the dataset. Accordingly, researchers can understand how well a variable can predict Alzheimer's disease and its overall link to the disorder.

There are various statistical techniques that are used to create predictive models. However, most statistical models in Alzheimer's disease research use regression analyses (e.g., see, Doody et al., 2010). In its simplest form, a regression uses the history of two variables to predict their future interaction (Ethington, Thomas, & Pike, 2002). There are many kinds of regressions that are equipped for different types of data. For example, there are regressions for linear and nonlinear data, single and multiple measures, and categorical and numerical variables (Schneider, Hommel, & Blettner, 2010; Yoo, Ramirez, & Liuzzi, 2014). Regressions are widely used throughout psychological research; however, most regressions are not feasible when using a large dataset. Accordingly, big data analysis commonly uses *multiple regressions* that specialize in analyzing multiple variables and modeling complex phenomena like Alzheimer's disease (Leech, Gliner, Morgan, & Harmon, 2003; Pandis, 2016; Passamonti et al., 2019). It is important to note that the mathematics behind multiple regressions is beyond the scope of this chapter.

The statistical modeling approach is currently the most common method of big data analysis in Alzheimer's disease research; however, these statistical techniques are rather simplistic compared to other big data methods (e.g., computational models and machine learning) (Gandomi & Haider, 2015). For example, statistical methods often struggle to process large amounts of data because they are limited by computational power or mathematical assumptions (Fan et al., 2014). Even statistical techniques that can process big data, such as multiple regressions, are restricted by their computational ability and underlying assumptions. For example, multiple regressions are open to suppression effects that sometimes bias variables (MacKinnon, Krull, & Lockwood, 2000). A suppressed variable is a poor predictive measure that is only included in a mixed model because other measures mask the poor predictors' effects (Beckstead, 2012). Suppressed variables are harmful to models as they provide no benefit and can confound results. These suppression effects can be overcome but at the cost of limiting variables and complicating analyses (Kraha, Turner, Nimon, Zientek, & Henson, 2012). Accordingly, researchers are progressively using computational models over traditional statistical methods to study Alzheimer's disease.

Computational modeling

Computational models use data to simulate and analyze complex systems (e.g., the brain, neurons). In big data research, scientists construct computational models to understand the components of a complex system. By understanding these components, computational models enable researchers to answer questions about underlying mechanism of a complex system. In Alzheimer's disease research, computational models can formulate more complex analyses than statistical models and can also create simulations of Alzheimer's disease (Saraceno, Musardo, Marcello, Pelucchi, & Diluca, 2013). Both statistical and computational modeling are mathematically similar; however, both forms of predictive modeling are distinguished by processing power, scientific approach, and technique (Hunt, Ropella, Park, & Engelberg, 2008). Statistical modeling is the mathematical analysis and interpretation of data while computational modeling is the technological analysis and simulation of data. Accordingly, computational models can perform the same tasks as statistical models but also excel in performing more complex analyses and simulations (Geerts, Hofmann-Apitius, Anastasio, & Brain Health Modeling Initiative, 2017). For example, computational models often use regression trees that provide more complex and detailed interpretations of data than standard statistical analyses (Kumar & Singh, 2017; Loh, 2011). Simulations excel at safely testing research hypotheses and exploring the mechanisms of Alzheimer's disease. For example, a model of disease biomarkers could be used to test new treatments for Alzheimer's disease (Hassan et al., 2018). Accordingly, computational models provide complex predictions of Alzheimer's disease and can enable fundamental exploratory research.

For example, Petrella, Hao, Rao, and Doraiswamy (2019) developed a computational model of biomarker cascade in Alzheimer's disease. The model consisted of Aβ, tau, cognitive decline, and neurodegenerative markers and was used to simulate Alzheimer's disease development in various populations (early-stage, autosomal dominant, and late-stage Alzheimer's disease). The computational model was constructed in MATLAB (a mathematical software) using the causal modeling approach and ordinary differential equations. Petrella et al. (2019) found that their cascade model accurately simulated cognitive decline as well as tau and Aβ progression in Alzheimer's disease. They also found that interactions between biomarkers accurately mimicked clinical trials run in the literature. They concluded that their model was a strong proof of concept and, with future refinement, could be used to assess the causal nature of Alzheimer's disease. However, Petrella et al. (2019) also stressed that the biomarker cascade model is only exploratory and is highly dependent on theoretical assumptions.

Computational models are extremely versatile and can range from single-cell simulations to massive population-based diagnostic models (Hassan et al., 2018). However, there have been varying levels of success with computational models in Alzheimer's disease research. Specifically, there has been great success formulating biochemical, single-cell, and system-level models (individual brain regions); however, current computational models often struggle to simulate complex phenomena like Alzheimer's disease development and progression (Hassan et al., 2018; Saraceno et al., 2013). These complex computational models are limited by our current understanding of Alzheimer's disease and the brain. Nonetheless, computational models are an ideal method for understanding Alzheimer's disease and are invaluable to dementia research. With further research and theory development, computational models will continue to forward our understanding of Alzheimer's disease.

The clinical applications of predictive models

Most of the predictive models discussed above focus on highly complex research topics such as the causes and progression of Alzheimer's disease; however, researchers are also developing predictive models that can directly assist with the clinical diagnosis and treatment of Alzheimer's disease. In this section, we explore how researchers are using electronic health records and patient monitoring technology to create predictive models for clinical use.

Electronic health records are government- (or institutional) run databases that contain citizens' medical data (Häyrinen, Saranto, & Nykänen, 2008). Electronic health records are often semi-automated systems that allow for the automated retrieval of health records from medical institutions and the manual input of information from doctors and patients. The popularity of electronic health records has increased in recent years as these databases are an efficient way of collecting medical information and streamlining healthcare practices. Importantly, scientists can access deidentified versions of electronic health records for research (Jensen, Jensen, & Brunak, 2012). In

Alzheimer's disease research, electronic health records are used to create predictive models that identify individuals at risk of Alzheimer's disease in the general population (Amra et al., 2017). Specifically, predictive analyses and machine learning are being used to analyze electronic health records and identify health patterns related to Alzheimer's disease development (e.g., cardiovascular health, stress, head trauma, age, and family history) (Perera et al., 2018). Accordingly, such predictive models could be used to identify individuals that are at risk of Alzheimer's disease development and inform early diagnosis and treatment plans.

Predictive analyses can also be used to monitor patients at risk of Alzheimer's disease and develop appropriate treatment plans. Specifically, technology, such as smartphone apps, is enabling researchers to easily and efficiently collect patient's data and monitor participants at risk of Alzheimer's disease (Sabbagh et al., 2020). These apps use games, quizzes, and puzzles to regularly collect participants' data for analysis (Clionsky & Clionsky, 2012). Using predictive analyses, the apps are then able to provide a summary of participants' health to doctors. In turn, doctors have more information available when deciding to diagnose or treat a patient with Alzheimer's disease (Clionsky & Clionsky, 2016; Ienca, Vayena, & Blasimme, 2018). For example, the diagnostic app CogniSense is used by clinicians to monitor patients with Alzheimer's disease and direct treatment (Quest Diagnostics, 2015). CogniSense is an iPad app that contains a series of questions and tasks that assess for the symptoms of Alzheimer's disease (e.g., memory loss). A full CogniSense assessment takes approximately 10 minutes and the test results are immediately given to the doctor. Accordingly, patient monitoring technology provides more information to doctors and can assist in the accelerated diagnosis and treatment of Alzheimer's disease.

However, it is important to note that there are significant ethical boundaries that must be overcome before predictive models can be applied to clinical practice. Namely, electronic health record and data monitoring technology must overcome problems with data security and consent to be clinically viable (Ienca et al., 2018). A history of data breaches and security concerns has significantly limited the adoption of electronic health records. For example, the Australian government's *My Health Records* has been plagued with security concerns, which has resulted in the mass exit of citizens from the program (McCall, 2018). If institutions want to use electronic health records, individuals must feel that their data are safe. Moreover, predictive models also have problems with consent and information accessibility (Arribas-Ayllon, 2011; Post et al., 1997). If an algorithm flags an individual as having Alzheimer's disease, whose job is it to tell the individual? Does the individual want to know? Can researchers make someone go to the doctor? Accordingly, the ethics of creating diagnostic and treatment-based programs on personal data can be obtrusive and messy in clinical practice. Clearly, current ethical quandaries will dictate the future application of big data to clinical practice.

Example: the Alzheimer's Disease Neuroimaging Initiative

Throughout this chapter, we have briefly discussed the ADNI as an example of big data collection in Alzheimer's disease research. Specifically, in section two we outlined how databases, such as ADNI, are instrumental in overcoming the logistical, temporal, and economic limitations that restrict standard Alzheimer's disease research (e.g., small laboratory-based studies). Furthermore, in section three, we discussed how databases enable the use of predictive models, which are better suited for Alzheimer's disease research compared to standard analytical measures (e.g., t-tests and simple linear regressions). In this section, we take an in-depth look at how ADNI collects and analyzes big data. We also aim to provide a brief example of the big data concepts discussed throughout this chapter. Accordingly, this section outlines ADNI's background, data collection methods, common analytical practices, and impact on Alzheimer's disease research.

The ADNI was formed in 2004 as both a publicly and privately funded entity that explores Alzheimer's disease biomarkers and disease prediction strategies. Most importantly for research, ADNI provides all its data to researchers for free as discussed above. ADNI has studied four cohorts over its history, with participants both carrying over from prior research and new participants being added at the beginning of each new initiative. ADNI studies typically last 5 years to allow for the longitudinal assessment of participants. In order, the ADNI cohorts thus far have been the ADNI1, the ADNIGO, the ADNI2, and the ADNI3 cohorts (Alzheimer's Disease Neuroimaging Initiative, 2017a). Upon the initiation of a new ADNI cohort, new biomarkers and research methodologies are added to keep the initiative up-to-date with the current Alzheimer's disease literature (Alzheimer's Disease Neuroimaging Initiative, 2017b). However, the iteration of initiatives sometimes makes it hard to examine measures across the ADNI cohorts.

The ADNI3 cohort is the most recent ongoing cohort, starting in 2016, and the ADNI2 cohort is the most recently completed study (Alzheimer's Disease Neuroimaging Initiative, 2016, p. 3). The ADNI2 cohort went for 5 years from 2011 to 2016 with participants' data recorded at annual or biannual intervals depending on the measures in question. The ADNI2 cohort consists of 700 participants from previous initiatives and 150 cognitively healthy controls, 100 early mild cognitive impairment, 150 late mild cognitive impairment, 150 Alzheimer's disease participants, and a new criterion of 107 participants with a significant memory concern (SMC) (Alzheimer's Disease Neuroimaging Initiative, 2017b). The SMC participants are, categorically, control subjects at a higher risk of disease conversion appearing on the Alzheimer's disease spectrum between healthy aging and MCI individuals.

Data types

Alzheimer's disease research is commonly categorized into the three fields of neuroimaging, biomarker, and psychological research. The ADNI database collects data from each of these fields to predict Alzheimer's disease progression and identify markers of the disease. ADNI chooses its markers and methods according to the surrounding literature; however, the popularity and accessibility of the ADNI database also significantly affect the prevalence of some measures in the literature.

Traditional *biomarker research* assesses biochemical changes in the brain that are associated with Alzheimer's disease. For example, Alzheimer's disease biomarker research commonly investigates genes, proteins, and peptides such as APOE4, tau, Aβ42, and presenilin-1 (Natelson Love et al., 2017; Sharma & Singh, 2016). The ADNI database collects a vast number of biomarkers for Alzheimer's disease research. Accordingly, ADNI collects blood, cerebral spinal fluid (CSF), and urine samples to study plasma, enzymes, proteins, amino acids, and genes as biomarkers of Alzheimer's disease. The use of traditional biomarkers in predictive models is debated as they are weak predictors of early-stage Alzheimer's disease; however, biomarkers are fundamental to current theories of Alzheimer's disease development such as the amyloid cascade hypothesis (Cui et al., 2011). Accordingly, biomarkers provide great insight into the progression of Alzheimer's disease and are instrumental in mixed predictor models of disease development.

Neuroimaging research commonly uses electromagnetic signals to noninvasively measure neurodegeneration and metabolic changes in Alzheimer's disease (e.g., MRI measures of hippocampal atrophy). Neuroimaging research is, in part, a form of biomarker research but is often separated from traditional biomarkers due to the size and unique methodology of the field. MRI and positron emission tomography (PET) are the most commonly used methods in neuroimaging research. In the ADNI database, MRI measures broadly assess brain atrophy, volume, and neuron connectivity while PET measures assess brain metabolism and Aβ pathology (Davatzikos, Bhatt, Shaw, Batmanghelich, & Trojanowski, 2011; Ewers et al., 2014; Landau et al., 2012). The use of neuroimaging technology is fundamental in all ADNI research as the technology is highly accurate when used to predict Alzheimer's disease. Accordingly, ADNI has helped cement MRI measures of brain volume and atrophy as some of the best predictors of Alzheimer's disease development and progression.

Psychological research focuses primarily on measuring cognitive and functional ability (e.g., motor skills and memory). Accordingly, pen-and-paper tests are commonly used to diagnose Alzheimer's disease and to monitor at-risk patients (e.g., MCI patients). ADNI currently uses 11 cognitive, and 10 functional and behavioral tests to monitor all Alzheimer's disease, mild cognitive impairment, and control participants. Cognitive tests can be further divided into three categories depending on the nature and use of the specific test (Brown, 2015). In Alzheimer's disease research, cognitive tests can broadly be categorized as the following: short questionnaires used for screening Alzheimer's disease (e.g., abbreviated mental test), highly specific tests used to discriminate between similar diseases such as vascular dementia and Alzheimer's disease (e.g., the Clock Drawing Test), and general multidomain tests commonly used for Alzheimer's disease diagnosis (e.g., the Mini-Mental State Examination) (Brown, 2015; Kato et al., 2013; Swain, O'Brien, & Nightingale, 1999). The use of cognitive tests is fundamental to the diagnosis, treatment, and study of Alzheimer's disease. Accordingly, ADNI research commonly uses psychological measures in predictive research and for clinical evaluations.

Mixed predictor models

The ADNI database has an immense amount of data that spans the various disciplines of Alzheimer's disease research. Accordingly, most ADNI studies often assess multiple variables at once in a mixed predictive model (e.g., episodic memory and hippocampal atrophy model) (Ramanan et al., 2012). Mixed predictive models are popular because they are an efficient way of studying multiple biomarkers at once (e.g., multiple regression). Studying multiple biomarkers at once is also important as Alzheimer's disease is seen as a multicausal disorder. The majority of ADNI studies combine MRI measures (e.g., brain volume or atrophy), memory scores, and Aβ and tau biomarkers (Davatzikos et al., 2011). For example, researchers have found strong relationships between medial temporal lobe atrophy (hippocampal and entorhinal volume/atrophy) and memory decline biomarkers (Ihara et al., 2018; Moradi et al., 2017). However, there is currently no standout mixed model that can predict Alzheimer's disease better than other good mixed predictive models. In reality, there are a handful of biomarkers that can predict Alzheimer's disease development with high accuracy depending on the stage of the disorder.

For example, a study by Huang et al. (2020) used the ADNI database and mixed predictive models to detect Alzheimer's disease development. Specifically, Huang et al. (2020) assessed the ability of cerebrospinal fluid, neuroimaging, and cognitive markers to predict Alzheimer's disease conversion in a sample of 290 mild cognitive impairment participants. Huang et al. (2020) found that a mixed predictive model outperformed all standalone predictors. Specifically, a mixed model consisting of genetic, Aβ, cerebral cortex, and Functional Activities Questionnaire measures was the best predictor of disease conversion compared to other models. Using their results, Huang et al. (2020) developed their mixed predictive model into a nomogram (scale) that clinicians can use to assess for Alzheimer's disease development. Accordingly, Huang et al. (2020) highlighted the strength of mixed predictive models and emphasized the importance of Alzheimer's disease biomarker in disease diagnosis.

ADNI data and predictive models have enabled researchers to make great breakthroughs in Alzheimer's disease research. In a review of all journal articles published using ADNI data, Weiner et al. (2015) outlined the breakthroughs of ADNI-based research. Specifically, Weiner et al. (2015) detailed that ADNI data have been instrumental in the development of neuroimaging technology for diagnostic purposes (e.g., MRI and PET), the discovery of early-stage Alzheimer's disease biomarkers (CSF biomarkers), and the investigation into potential predictors of Alzheimer's disease development (blood biomarkers) (Weiner et al., 2013, 2015). The initiative itself has also had a significant impact on research by pioneering big data techniques in the health sciences. For example, ADNI has inspired the development of many other Alzheimer's disease initiatives throughout the world such as the *Australian Imaging, Biomarker and Lifestyle*

Flagship Study of Ageing. Clearly, the ADNI database has helped to forward, inspire, and accelerate Alzheimer's disease research for the better.

Conclusion

Big data is actively changing the scope of Alzheimer's disease research. Through databases and predictive modeling, big data has managed to overcome many of the limitations of standard Alzheimer's disease research. Databases such as the ADNI have been instrumental in forwarding dementia research and enabling complex big data analyses (Weiner et al., 2015). Predictive modeling has also changed how researchers study Alzheimer's disease over time. Both computational and statistical methods have resulted in the formulation of complex mixed predictor models of Alzheimer's disease. With continued development these mixed predictor models will help clinicians to better diagnose, treat, and understand Alzheimer's disease. However, there are still significant hurdles that must be overcome for big data Alzheimer's disease research to progress. Specifically, researchers must overcome the ethical and organizational boundaries that limit big data research. Accordingly, it is essential that the researchers keep innovating and adapting big data techniques to Alzheimer's disease research. It is our hope that using big data, we as a scientific community can improve the lives of those with Alzheimer's disease and one day cure the disease.

References

Albert, M. S., DeKosky, S. T., Dickson, D., Dubois, B., Feldman, H. H., Fox, N. C., ... Phelps, C. H. (2011). The diagnosis of mild cognitive impairment due to Alzheimer's disease: Recommendations from the National Institute on Aging-Alzheimer's Association workgroups on diagnostic guidelines for Alzheimer's disease. *Alzheimer's & Dementia*, 7(3), 270–279. Available from https://doi.org/10.1016/j.jalz.2011.03.008.

Alzheimer's Association. (2018). 2018 Alzheimer's disease facts and figures. *Alzheimer's & Dementia*, *14*(3), 367–429.

Alzheimer's Disease International. (2019). *Dementia statistics*. Alzheimer's Disease International. https://www.alz.co.uk/research/statistics

Alzheimer's Disease Neuroimaging Initiative. (2016). *ADNI3 protocol*. http://adni.loni.usc.edu/wp-content/themes/freshnews-dev-v2/documents/clinical/ADNI3_Protocol.pdf

Alzheimer's Disease Neuroimaging Initiative. (2017a). *ADNI about*. Alzheimer's Disease Neuroimaging Initiative. http://adni.loni.usc.edu/about/

Alzheimer's Disease Neuroimaging Initiative. (2017b). *ADNI data types*. Alzheimer's Disease Neuroimaging Initiative. http://adni.loni.usc.edu/data-samples/data-types/

American Psychiatric Association. (2013). *Diagnostic and Statistical Manual of Mental Disorders* (5th Ed.). American Psychiatric Publishing.

Amra, S., O'Horo, J. C., Singh, T. D., Wilson, G. A., Kashyap, R., Petersen, R., ... Gajic, O. (2017). Derivation and validation of the automated search algorithms to identify cognitive impairment and dementia in electronic health records. *Journal of Critical Care*, *37*, 202–205. Available from https://doi.org/10.1016/j.jcrc.2016.09.026.

Arribas-Ayllon, M. (2011). The ethics of disclosing genetic diagnosis for Alzheimer's disease: Do we need a new paradigm? *British Medical Bulletin*, *100*(1), 7–21. Available from https://doi.org/10.1093/bmb/ldr023, http://orca.cf.ac.uk/view/cardiffauthors/A063035C.html.

Bartus, R. T., Dean, R. L., Beer, B., & Lippa, A. S. (1982). The cholinergic hypothesis of geriatric memory dysfunction. *Science (New York, N.Y.)*, *217*(4558), 408–414. Available from https://doi.org/10.1126/science.7046051.

Bastin, C., & Salmon, E. (2014). Early neuropsychological detection of Alzheimer's disease. *European Journal of Clinical Nutrition*, *68*(11), 1192–1199. Available from https://doi.org/10.1038/ejcn.2014.176.

Beckstead, J. W. (2012). Isolating and examining sources of suppression and multicollinearity in multiple linear regression. *Multivariate Behavioral Research*, *47*(2), 224–246. Available from https://doi.org/10.1080/00273171.2012.658331.

Bloom, G. S. (2014). Amyloid-β and tau: The trigger and bullet in Alzheimer disease pathogenesis. *JAMA Neurology*, *71*(4), 505–508. Available from https://doi.org/10.1001/jamaneurol.2013.5847.

Braak, H., & Del Tredici, K. (2011). The pathological process underlying Alzheimer's disease in individuals under thirty. *Acta Neuropathologica*, *121*(2), 171–181. Available from https://doi.org/10.1007/s00401-010-0789-4.

Bradway, C., Trotta, R., Bixby, M. B., McPartland, E., Wollman, M. C., Kapustka, H., ... Naylor, M. D. (2012). A qualitative analysis of an advanced practice nurse-directed transitional care model intervention. *The Gerontologist*, *52*(3), 394–407. Available from https://doi.org/10.1093/geront/gnr078.

Brown, J. (2015). The use and misuse of short cognitive tests in the diagnosis of dementia. *Journal of Neurology, Neurosurgery & Psychiatry*, *86*(6), 680–685. Available from https://doi.org/10.1136/jnnp-2014-309086.

Clionsky, M., & Clionsky, E. (2012). P4-277: The MOST-96120 iPad app improves PCP Alzheimer's disease screening. *Alzheimer's & Dementia*, *8*(4 Suppl. Part 20), S755–S756. Available from https://doi.org/10.1016/j.jalz.2013.08.058.

Clionsky, M., & Clionsky, E. (2016). Dementia and the brain-breathing connection. *Journal of Alzheimer's Disease & Parkinsonism*, *06*(06). Available from https://doi.org/10.4172/2161-0460.1000e135.

Cook, D. A., Andriole, D. A., Durning, S. J., Roberts, N. K., & Triola, M. M. (2010). Longitudinal research databases in medical education: Facilitating the study of educational outcomes over time and across institutions. *Academic Medicine*, *85*(8), 1340–1346. Available from https://doi.org/10.1097/ACM.0b013e3181e5c050.

Craig, L. A., Hong, N. S., & McDonald, R. J. (2011). Revisiting the cholinergic hypothesis in the development of Alzheimer's disease. *Neuroscience & Biobehavioral Reviews*, *35*(6), 1397–1409. Available from https://doi.org/10.1016/j.neubiorev.2011.03.001.

Csukly, G., Sirály, E., Fodor, Z., Horváth, A., Salacz, P., Hidasi, Z., ... Szabó, Á. (2016). The differentiation of amnestic type MCI from the non-amnestic types by structural MRI. *Frontiers in Aging Neuroscience*, *8*. Available from https://doi.org/10.3389/fnagi.2016.00052.

Cui, Y., Liu, B., Luo, S., Zhen, X., Fan, M., Liu, T., ... Alzheimer's Disease Neuroimaging Initiative. (2011). Identification of conversion from mild cognitive impairment to Alzheimer's disease using multivariate predictors. *PLoS One*, *6*(7), e21896. Available from https://doi.org/10.1371/journal.pone.0021896.

Cummings, J., Lai, T., Hemrungrojn, S., Mohandas, E., Yun Kim, S., Nair, G., & Dash, A. (2016). Role of donepezil in the management of neuropsychiatric symptoms in Alzheimer's disease and dementia with Lewy bodies. *CNS Neuroscience & Therapeutics*, *22*(3), 159–166. Available from https://doi.org/10.1111/cns.12484.

Davatzikos, C., Bhatt, P., Shaw, L. M., Batmanghelich, K. N., & Trojanowski, J. Q. (2011). Prediction of MCI to AD conversion, via MRI, CSF biomarkers, and pattern classification. *Neurobiology of Aging*, *32*(12), 2322. Available from https://doi.org/10.1016/j.neurobiolaging.2010.05.023, e19-27.

Doody, R. S., Pavlik, V., Massman, P., Rountree, S., Darby, E., & Chan, W. (2010). Predicting progression of Alzheimer's disease. *Alzheimer's Research & Therapy*, *2*(1), 2. Available from https://doi.org/10.1186/alzrt25.

Doraiswamy, P. M., Leon, J., Cummings, J. L., Marin, D., & Neumann, P. J. (2002). Prevalence and impact of medical comorbidity in Alzheimer's disease. *The Journals of Gerontology: Series A, Biological Sciences and Medical Sciences*, *57*(3), M173–M177. Available from https://doi.org/10.1093/gerona/57.3.M173.

Dubois, B., Feldman, H. H., Jacova, C., Cummings, J. L., DeKosky, S. T., Barberger-Gateau, P., ... Scheltens, P. (2010). Revising the definition of Alzheimer's disease: A new lexicon. *The Lancet Neurology*, *9*(11), 1118–1127. Available from https://doi.org/10.1016/S1474-4422(10)70223-4.

Eaton, M. L. (2001). Invariance in statistics. In N. J. Smelser, & P. B. Baltes (Eds.), *International encyclopedia of the social & behavioral sciences* (pp. 7893–7897). Pergamon. Available from https://doi.org/10.1016/B0-08-043076-7/00529-5.

Edvardsson, D., Fetherstonhaugh, D., & Nay, R. (2010). Promoting a continuation of self and normality: Person-centred care as described by people with dementia, their family members and aged care staff. *Journal of Clinical Nursing*, *19*(17–18), 2611–2618. Available from https://doi.org/10.1111/j.1365-2702.2009.03143.x.

Edvardsson, D., Winblad, B., & Sandman, P. (2008). Person-centred care of people with severe Alzheimer's disease: Current status and ways forward. *The Lancet Neurology*, 7(4), 362–367. Available from https://doi.org/10.1016/S1474-4422(08)70063-2.

El Haj, M., Antoine, P., Amouyel, P., Lambert, J.-C., Pasquier, F., & Kapogiannis, D. (2016). Apolipoprotein E (APOE) ε4 and episodic memory decline in Alzheimer's disease: A review. *Ageing Research Reviews*, *27*, 15–22. Available from https://doi.org/10.1016/j.arr.2016.02.002.

Ethington, C. A., Thomas, S. L., & Pike, G. R. (2002). Back to the basics: Regression as it should be. In J. C. Smart, & W. G. Tierney (Eds.), *Higher education: handbook of theory and research* (pp. 263–293). Netherlands: Springer. Available from https://doi.org/10.1007/978-94-010-0245-5_6.

Ewers, M., Brendel, M., Rizk-Jackson, A., Rominger, A., Bartenstein, P., Schuff, N., ... Alzheimer's Disease Neuroimaging Initiative (ADNI). (2014). Reduced FDG-PET brain metabolism and executive function predict clinical progression in elderly healthy subjects. *NeuroImage. Clinical*, *4*, 45–52. Available from https://doi.org/10.1016/j.nicl.2013.10.018.

Fan, J., Han, F., & Liu, H. (2014). Challenges of big data analysis. *National Science Review*, *1*(2), 293–314. Available from https://doi.org/10.1093/nsr/nwt032.

Feldman, H., Gauthier, S., Hecker, J., Vellas, B., Emir, B., Mastey, V., ... Group, T. D. M. S. I. (2003). Efficacy of donepezil on maintenance of activities of daily living in patients with moderate to severe Alzheimer's disease and the effect on caregiver burden. *Journal of the American Geriatrics Society*, *51*(6), 737–744. Available from https://doi.org/10.1046/j.1365-2389.2003.51260.x.

Forlenza, O. V., Diniz, B. S., Talib, L. L., Radanovic, M., Yassuda, M. S., Ojopi, E. B., & Gattaz, W. F. (2010). Clinical and biological predictors of Alzheimer's disease in patients with amnestic mild cognitive impairment. *Revista Brasileira De Psiquiatria*, *32*(3), 216–222.

Gandomi, A., & Haider, M. (2015). Beyond the hype: Big data concepts, methods, and analytics. *International Journal of Information Management*, *35*(2), 137–144. Available from https://doi.org/10.1016/j.ijinfomgt.2014.10.007.
Geerts, H., Dacks, P. A., Devanarayan, V., Haas, M., Khachaturian, Z. S., Gordon, M. F., ... Stephenson, D. (2016). Big data to smart data in Alzheimer's disease: The brain health modeling initiative to foster actionable knowledge. *Alzheimer's & Dementia*, *12*(9), 1014–1021. Available from https://doi.org/10.1016/j.jalz.2016.04.008.
Geerts, H., Hofmann-Apitius, M., Anastasio, T. J., & Brain Health Modeling Initiative. (2017). Knowledge-driven computational modeling in Alzheimer's disease research: Current state and future trends. *Alzheimer's & Dementia*, *13*(11), 1292–1302. Available from https://doi.org/10.1016/j.jalz.2017.08.011.
Hassan, M., Abbas, Q., Seo, S.-Y., Shahzadi, S., Ashwal, H. A., Zaki, N., ... Moustafa, A. A. (2018). Computational modeling and biomarker studies of pharmacological treatment of Alzheimer's disease. *Molecular Medicine Reports*, *18*(1), 639–655. Available from https://doi.org/10.3892/mmr.2018.9044.
Häyrinen, K., Saranto, K., & Nykänen, P. (2008). Definition, structure, content, use and impacts of electronic health records: A review of the research literature. *International Journal of Medical Informatics*, 77(5), 291–304. Available from https://doi.org/10.1016/j.ijmedinf.2007.09.001.
Huang, K., Lin, Y., Yang, L., Wang, Y., Cai, S., Pang, L., ... Huang, L. (2020). A multi-predictor model to predict the conversion of mild cognitive impairment to Alzheimer's disease by using a predictive nomogram. *Neuropsychopharmacology: Official Publication of the American College of Neuropsychopharmacology*, *45*(2), 358–366. Available from https://doi.org/10.1038/s41386-019-0551-0.
Hunt, C. A., Ropella, G. E. P., Park, S., & Engelberg, J. (2008). Dichotomies between computational and mathematical models. *Nature Biotechnology*, *26*(7), 737–738. Available from https://doi.org/10.1038/nbt0708-737.
Ienca, M., Vayena, E., & Blasimme, A. (2018). Big data and dementia: Charting the route ahead for research, ethics, and policy. *Frontiers in Medicine*, *5*. Available from https://doi.org/10.3389/fmed.2018.00013.
Ihara, R., Iwata, A., Suzuki, K., Ikeuchi, T., Kuwano, R., & Iwatsubo, T. (2018). Clinical and cognitive characteristics of preclinical Alzheimer's disease in the Japanese Alzheimer's disease neuroimaging initiative cohort. *Alzheimer's & Dementia: Translational Research & Clinical Interventions*, *4*, 645–651. Available from https://doi.org/10.1016/j.trci.2018.10.004.
Iqbal, K., Liu, F., Gong, C.-X., & Grundke-Iqbal, I. (2010). Tau in Alzheimer disease and related tauopathies. *Current Alzheimer Research*, 7(8), 656–664.
Jack, C. R., Albert, M., Knopman, D. S., McKhann, G. M., Sperling, R. A., Carillo, M., ... Phelps, C. H. (2011). Introduction to revised criteria for the diagnosis of Alzheimer's disease: National Institute on Aging and the Alzheimer Association Workgroups. *Alzheimer's & Dementia*, 7(3), 257–262. Available from https://doi.org/10.1016/j.jalz.2011.03.004.
Jack, C. R., Knopman, D. S., Jagust, W. J., Shaw, L. M., Aisen, P. S., Weiner, M. W., ... Trojanowski, J. Q. (2010). Hypothetical model of dynamic biomarkers of the Alzheimer's pathological cascade. *Lancet Neurology*, *9*(1), 119. Available from https://doi.org/10.1016/S1474-4422(09)70299-6.
Jensen, P. B., Jensen, L. J., & Brunak, S. (2012). Mining electronic health records: Towards better research applications and clinical care. *Nature Reviews. Genetics*, *13*(6), 395–405. Available from https://doi.org/10.1038/nrg3208.
Kalat, J. W. (2018). *Biological psychology* (13th ed). Cengage.
Kato, Y., Narumoto, J., Matsuoka, T., Okamura, A., Koumi, H., Kishikawa, Y., ... Fukui, K. (2013). Diagnostic performance of a combination of Mini-Mental State Examination and Clock Drawing Test in detecting Alzheimer's disease. *Neuropsychiatric Disease and Treatment*, *9*, 581–586. Available from https://doi.org/10.2147/NDT.S42209.

Kraha, A., Turner, H., Nimon, K., Zientek, L., & Henson, R. (2012). Tools to support interpreting multiple regression in the face of multicollinearity. *Frontiers in Psychology*, *3*. Available from https://doi.org/10.3389/fpsyg.2012.00044.

Kumar, A., & Singh, T. R. (2017). A new decision tree to solve the puzzle of Alzheimer's disease pathogenesis through standard diagnosis scoring system. *Interdisciplinary Sciences, Computational Life Sciences*, *9*(1), 107–115. Available from https://doi.org/10.1007/s12539-016-0144-0.

Landau, S. M., Mintun, M. A., Joshi, A. D., Koeppe, R. A., Petersen, R. C., Aisen, P. S., ... Jagust, W. J. (2012). Amyloid deposition, hypometabolism, and longitudinal cognitive decline. *Annals of Neurology*, *72*(4), 578–586. Available from https://doi.org/10.1002/ana.23650.

Larner, A. J. (2019). Evaluating cognitive screening instruments with the "likelihood to be diagnosed or misdiagnosed" measure. *International Journal of Clinical Practice*, *73*(2), e13265. Available from https://doi.org/10.1111/ijcp.13265.

Leech, N. L., Gliner, J. A., Morgan, G. A., & Harmon, R. J. (2003). Use and interpretation of multiple regression. *Journal of the American Academy of Child & Adolescent Psychiatry*, *42*(6), 738–740. Available from https://doi.org/10.1097/01.CHI.0000046845.56865.22.

Loh, W.-Y. (2011). Classification and regression trees. *WIREs Data Mining and Knowledge Discovery*, *1*(1), 14–23. Available from https://doi.org/10.1002/widm.8.

Ma, Y., Ji, J., Li, G., Yang, S., & Pan, S. (2018). Effects of donepezil on cognitive functions and the expression level of β-amyloid in peripheral blood of patients with Alzheimer's disease. *Experimental and Therapeutic Medicine*, *15*(2), 1875–1878.

MacKinnon, D. P., Krull, J. L., & Lockwood, C. M. (2000). Equivalence of the mediation, confounding and suppression effect. *Prevention Science*, *1*(4), 173–181. Available from https://doi.org/10.1023/A:1026595011371.

Martone, M. E., Garcia-Castro, A., & VandenBos, G. R. (2018). Data sharing in psychology. *The American Psychologist*, *73*(2), 111–125. Available from https://doi.org/10.1037/amp0000242.

Mayer, D. G., & Butler, D. G. (1993). Statistical validation. *Ecological Modelling*, *68*(1), 21–32. Available from https://doi.org/10.1016/0304-3800(93)90105-2.

McCall, C. (2018). Opt-out digital health records cause debate in Australia. *The Lancet*, *392*(10145), 372. Available from https://doi.org/10.1016/S0140-6736(18)31726-4.

McKhann, G. M., Knopman, D. S., Chertkow, H., Hyman, B. T., Jack, C. R., Jr., Kawas, C. H., ... Phelps, C. H. (2011). The diagnosis of dementia due to Alzheimer's disease: Recommendations from the National Institute on Aging-Alzheimer's Association workgroups on diagnostic guidelines for Alzheimer's disease. *Alzheimer's & Dementia*, 7(3), 263–269. Available from https://doi.org/10.1016/j.jalz.2011.03.005.

Moradi, E., Hallikainen, I., Hänninen, T., Tohka, J., & Alzheimer's Disease Neuroimaging Initiative. (2017). Rey's Auditory Verbal Learning Test scores can be predicted from whole brain MRI in Alzheimer's disease. *NeuroImage. Clinical*, *13*, 415–427. Available from https://doi.org/10.1016/j.nicl.2016.12.011.

Mueller, S. G., Weiner, M. W., Thal, L. J., Petersen, R. C., Jack, C., Jagust, W., ... Beckett, L. (2005). The Alzheimer's Disease Neuroimaging Initiative. *Neuroimaging Clinics of North America*, *15*(4), 869. Available from https://doi.org/10.1016/j.nic.2005.09.008, xii.

Murphy, M. P., & LeVine, H. (2010). Alzheimer's disease and the β-amyloid peptide. *Journal of Alzheimer's Disease: JAD*, *19*(1), 311. Available from https://doi.org/10.3233/JAD-2010-1221.

Nakagawa, S., & Freckleton, R. P. (2008). Missing inaction: The dangers of ignoring missing data. *Trends in Ecology & Evolution*, *23*(11), 592–596. Available from https://doi.org/10.1016/j.tree.2008.06.014.

Natelson Love, M., Clark, D. G., Cochran, J. N., Den Beste, K. A., Geldmacher, D. S., Benzinger, T. L., ... Roberson, E. D. (2017). Clinical, imaging, pathological, and biochemical characterization of a novel presenilin 1 mutation (N135Y) causing Alzheimer's disease. *Neurobiology of Aging, 49*, 216.e7–216.e13. Available from https://doi.org/10.1016/j.neurobiolaging.2016.09.020.

Olfson, M., Wall, M. M., & Blanco, C. (2017). Incentivizing data sharing and collaboration in medical research—The S-index. *JAMA Psychiatry, 74*(1), 5–6. Available from https://doi.org/10.1001/jamapsychiatry.2016.2610.

Pandis, N. (2016). Multiple linear regression analysis. *American Journal of Orthodontics and Dentofacial Orthopedics, 149*(4), 581. Available from https://doi.org/10.1016/j.ajodo.2016.01.012.

Papale, D., Agarwal, D. A., Baldocchi, D., Cook, R. B., Fisher, J. B., & van Ingen, C. (2012). Database maintenance, data sharing policy, collaboration. In M. Aubinet, T. Vesala, & D. Papale (Eds.), *Eddy covariance: a practical guide to measurement and data analysis* (pp. 399–424). Netherlands: Springer. Available from https://doi.org/10.1007/978-94-007-2351-1_17.

Passamonti, L., Tsvetanov, K. A., Jones, P. S., Bevan-Jones, W. R., Arnold, R., Borchert, R. J., ... Rowe, J. B. (2019). Neuroinflammation and functional connectivity in Alzheimer's disease: Interactive influences on cognitive performance. *Journal of Neuroscience, 39*(36), 7218–7226. Available from https://doi.org/10.1523/JNEUROSCI.2574-18.2019.

Pearlson, G. (2009). Multisite collaborations and large databases in psychiatric neuroimaging: Advantages, problems, and challenges. *Schizophrenia Bulletin, 35*(1), 1–2. Available from https://doi.org/10.1093/schbul/sbn166.

Perera, G., Pedersen, L., Ansel, D., Alexander, M., Arrighi, H. M., Avillach, P., ... Stewart, R. (2018). Dementia prevalence and incidence in a federation of European Electronic Health Record databases: The European Medical Informatics Framework resource. *Alzheimer's & Dementia, 14*(2), 130–139. Available from https://doi.org/10.1016/j.jalz.2017.06.2270.

Perry, R. J., Watson, P., & Hodges, J. R. (2000). The nature and staging of attention dysfunction in early (minimal and mild) Alzheimer's disease: Relationship to episodic and semantic memory impairment. *Neuropsychologia, 38*(3), 252–271. Available from https://doi.org/10.1016/S0028-3932(99)00079-2.

Petrella, J. R., Hao, W., Rao, A., & Doraiswamy, P. M. (2019). Computational causal modeling of the dynamic biomarker cascade in Alzheimer's disease. *Computational and Mathematical Methods in Medicine, 2019*, 6216530, Hindawi. Available from https://doi.org/10.1155/2019/6216530.

Post, S. G., Whitehouse, P. J., Binstock, R. H., Bird, T. D., Eckert, S. K., Farrer, L. A., ... Zinn, A. B. (1997). The clinical introduction of genetic testing for Alzheimer disease: An ethical perspective. *JAMA: The Journal of the American Medical Association, 277*(10), 832–836. Available from https://doi.org/10.1001/jama.1997.03540340066035.

Quest Diagnostics. (2015). *CogniSense^{TM} iPad App for cognitive impairment screening*. CogniSense. http://www.questcognisense.com/

Ramanan, V., Shen, L., Risacher, S., Kim, S., Boddu, M., West, J., ... Saykin, A. (2012). Hippocampal subfield atrophy in the Alzheimer's Disease Neuroimaging Initiative (ADNI) cohort: Relationships to diagnosis and memory impairment. *Alzheimer's & Dementia, 8*(4 Suppl.), P533–P534. Available from https://doi.org/10.1016/j.jalz.2012.05.1434.

Roberts, R., Geda, Y., Knopman, D., Cha, R., Pankratz, V., Boeve, B., ... Rocca, W. (2008). The Mayo Clinic Study of Aging: Design and sampling, participation, baseline measures and sample characteristics. *Neuroepidemiology, 30*(1), 58–69. Available from https://doi.org/10.1159/000115751.

Rodrigue, K. M., Kennedy, K. M., Devous, M. D., Rieck, J. R., Hebrank, A. C., Diaz-Arrastia, R., ... Park, D. C. (2012). β-Amyloid burden in healthy aging. *Neurology, 78*(6), 387–395. Available from https://doi.org/10.1212/WNL.0b013e318245d295.

M.N. Sabbagh, M. Boada, S. Borson, M. Chilukuri, B. Dubois, J. Ingram, ... H. Hampel. (2020). *Early detection of mild cognitive impairment (MCI) in primary care*. https://doi.org/10.14283/JPAD.2020.21

Saraceno, C., Musardo, S., Marcello, E., Pelucchi, S., & Diluca, M. (2013). Modeling Alzheimer's disease: From past to future. *Frontiers in Pharmacology, 4*. Available from https://doi.org/10.3389/fphar.2013.00077.

Schmitt, C. P., & Burchinal, M. (2011). Data management practices for collaborative research. *Frontiers in Psychiatry, 2*. Available from https://doi.org/10.3389/fpsyt.2011.00047.

Schneider, A., Hommel, G., & Blettner, M. (2010). Linear regression analysis. *Deutsches Ärzteblatt International, 107*(44), 776–782. Available from https://doi.org/10.3238/arztebl.2010.0776.

Sharma, N., & Singh, A. N. (2016). Exploring biomarkers for Alzheimer's disease. *Journal of Clinical and Diagnostic Research: JCDR, 10*(7), KE01–KE06. Available from https://doi.org/10.7860/JCDR/2016/18828.8166.

Small, G., & Bullock, R. (2011). Defining optimal treatment with cholinesterase inhibitors in Alzheimer's disease. *Alzheimer's & Dementia*, 7(2), 177–184. Available from https://doi.org/10.1016/j.jalz.2010.03.016.

Stites, S. D., Milne, R., & Karlawish, J. (2018). Advances in Alzheimer's imaging are changing the experience of Alzheimer's disease. *Alzheimer's & Dementia: Diagnosis, Assessment & Disease Monitoring, 10*, 285–300. Available from https://doi.org/10.1016/j.dadm.2018.02.006.

Swain, D. G., O'Brien, A. G., & Nightingale, P. G. (1999). Cognitive assessment in elderly patients admitted to hospital: The relationship between the Abbreviated Mental Test and the Mini-Mental State Examination. *Clinical Rehabilitation, 13*(6), 503–508. Available from https://doi.org/10.1191/026921599670895633.

Tan, C.-C., Yu, J.-T., Wang, H.-F., Tan, M.-S., Meng, X.-F., Wang, C., ... Tan, L. (2014). Efficacy and safety of donepezil, galantamine, rivastigmine, and memantine for the treatment of Alzheimer's disease: A systematic review and meta-analysis. *Journal of Alzheimer's Disease, 41*(2), 615–631. Available from https://doi.org/10.3233/JAD-132690.

Toga, A. W., Bhatt, P., & Ashish, N. (2016). Global data sharing in Alzheimer's disease research. *Alzheimer Disease and Associated Disorders, 30*(2), 160–168. Available from https://doi.org/10.1097/WAD.0000000000000121.

Vikström, P., Edvinsson, S., & Brändström, A. (2002). Longitudinal databases - Sources for analyzing the life-course: Characteristics, difficulties and possibilities. *History and Computing, 14*(1–2), 109–128. Available from https://doi.org/10.3366/hac.2002.14.1-2.109.

Wada, H., Nakajoh, K., Satoh-Nakagawa, T., Suzuki, T., Ohrui, T., Arai, H., & Sasaki, H. (2001). Risk factors of aspiration pneumonia in Alzheimer's disease patients. *Gerontology, 47*(5), 271–276. Available from https://doi.org/10.1159/000052811.

Watson, J. L., Ryan, L., Silverberg, N., Cahan, V., & Bernard, M. A. (2014). Obstacles and opportunities in Alzheimer's clinical trial recruitment. *Health Affairs (Project Hope), 33*(4), 574–579. Available from https://doi.org/10.1377/hlthaff.2013.1314.

Weiner, M. W., Veitch, D. P., Aisen, P. S., Beckett, L. A., Cairns, N. J., Cedarbaum, J., ... Alzheimer's Disease Neuroimaging Initiative. (2015). 2014 Update of the Alzheimer's Disease Neuroimaging Initiative: A review of papers published since its inception. *Alzheimer's & Dementia, 11*(6). Available from https://doi.org/10.1016/j.jalz.2014.11.001, e1-120.

Weiner, M. W., Veitch, D. P., Aisen, P. S., Beckett, L. A., Cairns, N. J., Green, R. C., ... Alzheimer's Disease Neuroimaging Initiative. (2013). The Alzheimer's Disease Neuroimaging Initiative: A review of papers published since its inception. *Alzheimer's & Dementia*, *9*(5), e111–e194. Available from https://doi.org/10.1016/j.jalz.2013.05.1769.

World Health Organization. (2017). *Global action plan on the public health response to dementia 2017—2025.* https://www.who.int/mental_health/neurology/dementia/action_plan_2017_2025/en/

World Health Organization. (2018). *Global Health Estimates 2016: Deaths by cause, age, sex, by country and by region, 2000-2016.* https://www.who.int/gho/mortality_burden_disease/causes_death/top_10/en/

Yankner, B. A., Duffy, L. K., & Kirschner, D. A. (1990). Neurotrophic and neurotoxic effects of amyloid beta protein: Reversal by tachykinin neuropeptides. *Science (New York, N.Y.)*, *250*(4978), 279–282. Available from https://doi.org/10.1126/science.2218531.

Yiannopoulou, K. G., & Papageorgiou, S. G. (2013). Current and future treatments for Alzheimer's disease. *Therapeutic Advances in Neurological Disorders*, *6*(1), 19–33. Available from https://doi.org/10.1177/1756285612461679.

Yoo, C., Ramirez, L., & Liuzzi, J. (2014). Big data analysis using modern statistical and machine learning methods in medicine. *International Neurourology Journal*, *18*(2), 50–57. Available from https://doi.org/10.5213/inj.2014.18.2.50.

CHAPTER 8

Operational aspects of deep learning solutions for Alzheimer's disease

Samuel L. Warren[1], Ahmed A. Moustafa[2,3,4] and Dustin van der Haar[5]

[1]Psychological Science, School of Psychology, Western Sydney University, Sydney, NSW, Australia
[2]School of Psychology, Western Sydney University, Sydney, NSW, Australia
[3]MARCS Institute for Brain and Behaviour, Western Sydney University, Sydney, NSW, Australia
[4]Department of Human Anatomy and Physiology, The Faculty of Health Sciences, University of Johannesburg, Johannesburg, South Africa
[5]Academy of Computer Science and Software Engineering at the University of Johannesburg, South Africa

Introduction

Alzheimer's disease (AD) is the most common form of dementia, characterized by severe cognitive decline and neurodegeneration (Dubois et al., 2007). While there has been extensive research on the diagnosis and treatment of AD in recent decades, the capacity of these combative measures is still severely limited. For example, modern AD diagnoses are prone to misdiagnosing different forms of dementia (e.g., diagnosing vascular dementia as AD) and are unable to detect the early (and potentially reversible) stages of the disease (Arevalo-Rodriguez et al., 2015; James et al., 2020). In turn, there is currently no cure for AD and the disease remains one of the leading causes of death worldwide (WHO, 2019). Consequentially, there is a critical need for diagnostic methods that can diagnose AD at an early and curable stage.

Historically, researchers have used statistical models (e.g., linear regressions) to predict the early stages of AD and, thus identify the disease at a treatable stage (e.g., see Xu, Chen, Zhao, Li, & Guo, 2018). However, it is thought that these models are often too simple to fully encapsulate the complex, multicausal nature of AD. Accordingly, neuroscientific research is increasingly transitioning from statistical to computational diagnostic methods (e.g., machine and deep learning). This rise of computer-assisted diagnosis (CAD) can be attributed to multiple technological and scientific advancements. For example, the growth of CAD can be linked to advancements in artificial intelligence (AI), computer hardware (e.g., computer processing

Alzheimer's Disease
DOI: https://doi.org/10.1016/B978-0-12-821334-6.00002-8

units), data sharing, open-source software, neuroimaging technology [e.g., magnetic resonance imaging (MRI)], and the discovery of AD biomarkers (biological indicators of AD) (Ausó, Gómez-Vicente, & Esquiva, 2020; Subasi, 2020; Wang & Raj, 2017). Through these advancements, CAD methods have been able to overcome some of the limitations of prior statistical approaches (e.g., simplistic representation of AD, mediocre diagnostic accuracy).

In AD research, CAD methods are generally categorized into the two broad fields of machine learning and deep learning. Both machine learning and deep learning are types of AI that seek to computationally process and categorize data. While these AI methods have the same goal, they are differentiated by their complexity, methodology, and independence. Machine learning methods are the oldest and most simplistic of the two categories. These traditional CAD methods work using multiple statistical tests to identify, extract, and classify data (Tanveer et al., 2020). However, traditional CAD models do have some limitations. For example, traditional CAD models are time-consuming, require supervision, and lack application to clinical practice (Pellegrini et al., 2018). Alternatively, deep learning methods are newer techniques that use neural networks to imitate human learning (Jo, Nho, & Saykin, 2019). Deep learning methods are superior to traditional CAD methods as they are more efficient, can run unsupervised, and can classify AD with higher accuracy (Ebrahimighahnavieh, Luo, & Chiong, 2020). Accordingly, deep learning-based CAD systems are revolutionizing AD research.

In this chapter, we describe the application and operationalization of deep learning methods in AD research. Specifically, we discuss the following topics: (1) traditional CAD methods and their lessons for deep learning-based CAD systems; (2) the application of deep learning methods to AD research; and (3) the advantages and limitations of deep learning methods in AD research. In turn, we aim to equip readers with an in-depth understanding of deep learning-based CAD systems and promote further research into the diagnosis and treatment of AD.

Traditional Alzheimer's CAD method trends

Introduction

Traditional CAD methods use statistical and machine learning techniques to detect AD (Tanveer et al., 2020). These traditional CAD methods are popular in research due to their ability to analyze large datasets and accurately detect AD. Accordingly, traditional CAD models are commonly used to streamline data analysis and inform AD diagnoses (i.e., provide a second opinion) (Shatte, Hutchinson, & Teague, 2019). However, traditional CAD methods do have some limitations (e.g., lack of autonomy and generalizability) and, thus lack clinical application (Pellegrini et al., 2018; Miotto, Wang, Wang, Jiang, & Dudley, 2018).

Consequentially, traditional CAD models are being superseded by deep learning methods in AD research. These deep learning models are popular because they are highly accurate and can overcome many of the limitations of traditional CAD models (Ebrahimighahnavieh et al., 2020). For example, deep learning models can run unsupervised, unlike most traditional methods (Bi et al., 2020). Nonetheless, traditional CAD models are useful and have many similarities in common with deep learning methods. Thus it is important to review traditional CAD methods to inform emerging deep learning models. In this section, we discuss the common methods and stages of traditional CAD research.

CAD methods in AD research

A traditional CAD model involves three stages known as data preprocessing, feature extraction, and classification (see Fig. 8.1) (Raghavendra, Acharya, & Adeli, 2019). The aims of these stages are as follows: data preprocessing aims to clean and normalize a dataset; feature extraction aims to identify and isolate variables of interest; and classification seeks to train, test, and evaluate a diagnostic model (Karami, Nittari, & Amenta, 2019). Each of these stages uses different statistical and computational methods to process AD data. Of these methods, a researcher will often select specific analyses depending on the types of data collected. For example, the statistical transformations used to preprocess brain images (e.g., MRI) are different to the methods used to study electrical brain signals (e.g., EEG) (Hulbert & Adeli, 2013).

Some methodologies are also preferred due to the popularity of technology and data collection methods. For example, in traditional CAD

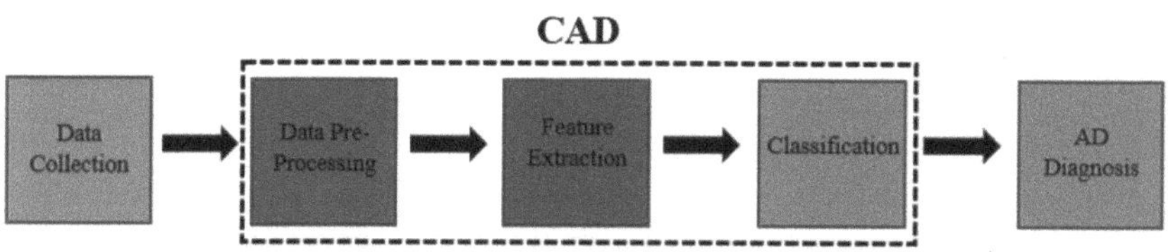

Figure 8.1 The general stages of a traditional CAD.

research, neuroimaging biomarkers are commonly used to diagnose AD (e.g., MRI brain volume measures). These imaging methods are favored because they can directly observe AD, are noninvasive, and have high diagnostic accuracy (Karami et al., 2019; Pellegrini et al., 2018). Accordingly, most CAD methods in AD research focus on processing and analyzing neuroimaging data. A handful of studies use other AD measures to perform diagnoses (e.g., memory tests); however, these nonimaging methodologies are rare (Segovia et al., 2014). Consequentially, in this section, we outline the three stages of traditional CAD research with an emphasis on neuroimaging methods.

Data preprocessing

Data preprocessing is the first stage of traditional CAD methodology. During the stage of preprocessing, data are cleaned, normalized, and transformed for feature extraction (Valenzuela, Jiang, Carrillo, & Rojas, 2018). The data preprocessing stage is the foundation of a CAD model and is critical to a model's classification accuracy. Accordingly, data must be appropriately obtained, cleaned, and managed during preprocessing. In AD research, data preprocessing usually involves statistical or computational transformations that simplify and standardize neuroimages.

Specifically, statistical parametric mapping (SPM) and voxel-based morphometry (VBM) are the most popular techniques used (Dessouky, Elrashidy, Taha, & Abdelkader, 2016). SPM works by statistically analyzing and comparing groups of images to highlight neurological differences (Chaves et al., 2009). SPM also typically involves the alignment, smoothing, and univariate analysis of images. Similarly, VBM uses SPM techniques to compare images, yet differs by using computational comparisons of brain-tissue voxels to highlight neurological differences (a voxel is a computational pixel or 3D representation of the brain) (Zhang et al., 2019). Both SPM and VBM techniques have corresponding software that can streamline and semiautomate data preprocessing (e.g., SPM12 and FreeSurfer).

Feature extraction

Feature extraction is the second stage of CAD methodology. During the feature extraction stage variables of interest are identified, isolated, and processed (Segovia et al., 2014). In turn, the feature stage aims to detect and refine variables for future classification. The feature stage is typically split into three steps known as feature selection, extraction, and reduction (also known as dimensional reduction) (Khalid, Khalil, & Nasreen, 2014). First, feature selection is the process of identifying and categorizing regions or variables of interest. Next, feature extraction involves the isolation of the region or variable of interest from the surrounding image or data. Lastly, feature reduction involves the compression of data to reduce the processing load during classification (feature reduction is mostly used with imaging data).

Some common methods of processing features are discrete wave transformations, Fourier analysis, partial least squares, and principal component analysis (PCA) (Dessouky et al., 2016). These methods mostly use mathematical transformations or correlations to process features. For example, a PCA is a statistical test that identifies and groups similar variables according to their relationship to each other (Abdi & Williams, 2010; Shaikh & Ali, 2019). Selecting the correct feature extraction technique is pivotal to the clarity and accuracy of a CAD model. Accordingly, the ideal methods for processing features are often determined by prior research and the tests fit the dataset. It is also common for studies to compare multiple feature extraction techniques to determine which methods result in the most accurate CAD models (Dessouky et al., 2016).

Classification

Classification is the third and final stage of traditional CAD methodology. During the classification stage, diagnosis models are developed through the iterative training, testing, and comparison of different CAD methods (Ruiz, Ramírez, Górriz, Casillas, & Initiative, 2018). These steps of training, testing, and evaluation are all interlinked and are constantly refined to construct a highly accurate CAD model. During the initial steps of classification, a dataset is partitioned with separate parts used to train and test the diagnostic model. There are multiple machine learning methods that are used to train and test a classification model such as, support vector machines, random forests, and logistic regressions (Karami et al., 2019; Podgorelec, 2012; Raghavendra et al., 2019). Each classification model works using similar categorization techniques that seek to discriminate between diagnostic groups (e.g., AD and control participants). For example, support vector machines work by plotting data (e.g., brain volume) on a plane and determining diagnoses according to the relationship between data points (in reference to the training categories) (Raghavendra et al., 2019; Valenzuela et al., 2018).

Cross-validation techniques commonly determine the specific machine learning method used to train and classify the data. Cross-validation techniques are ongoing throughout classification and seek to evaluate a model by comparing different training/testing criteria and diagnostic methods (Shakarami, Tarrah, & Mahdavi-Hormat, 2020). This evaluation of classification techniques is commonly done by partitioning the dataset and comparing the results of different methodologies (e.g., support vector machines versus logistic regression). For example, 10-fold cross-validation separates the dataset into 10 even groups with every group alternating as a training or testing group. A series of different classification methods are also tested and compared. The results indicate which training/testing combination and classification methods result in the highest diagnostic accuracy (Lahmiri & Shmuel, 2019).

External influences on the application of the model

It is important to note that the stages of data preprocessing, feature extraction, and classification are not the only influences on the accuracy and application of a CAD model. External influences such as data quality and clinical opinion (e.g., a clinician's willingness to use a model) can also affect the use of CAD models. Regarding data collection, it is important to understand that the quality and clarity of a dataset will directly influence the success of a CAD model. A weak dataset will result in a poor CAD model regardless of data preprocessing and feature extraction techniques. Moreover, a small sample will result in an unrepresentative model and poor classification.

Accordingly, it is important that data are of high quality (e.g., very little missing data) and is representative of individuals with AD. This need for representative and high-quality data can be costly. However, most researchers overcome these boundaries by using collaborative databases. Datasets should include the collection of high-quality data thereby enabling most CAD research (see, Bi et al., 2020; Ding et al., 2018). The majority of AD data is acquired from databases such as the *Alzheimer's Disease Neuroimaging Initiative; Australian Imaging, Biomarker and Lifestyle Flagship Study of Ageing; UK Biobank and The Global Alzheimer's Association Interactive Network.*

It is also important to understand that clinicians influence the accuracy and application of CAD models. Traditional CAD models can only run when supervised, and thus rely on clinicians to verify and perform diagnoses. Accordingly, CAD methods' success depends on a clinician's ability to trust, use, and interpret a model (e.g., Krupinski, 2020). If a clinician feels that a model is not accurate or time-efficient, they will not use it. Consequentially, CAD models must be optimized so that clinicians find them reliable, user friendly, and easily interpretable (Miotto et al., 2018). Researchers should consider including clinicians in the creation of CAD models. Researchers should also ensure that CAD models are reliable in clinical samples and that adequate training resources are available. By managing this relationship between CAD models and clinicians, researchers can improve CAD accuracy and application.

Conclusion

In this section, we outlined the common stages of traditional CAD as well as external factors that can influence diagnostic models. We also discussed common methods that are used to process, extract, and classify data for diagnosis. While some of these methods may not directly transfer to deep learning models, it is important to understand that the overarching topics do. Traditional CAD does have its limitations concerning the generalizability and supervision of diagnostic models; however, the intricate and iterative nature of traditional CAD methods are admirable. Accordingly, emerging CAD methods should learn from the iterative and systematic nature of traditional CAD. Deep learning models should also consider the external factors that influence the application of CAD. If these barriers can be overcome, CAD could revolutionize AD diagnoses.

Alzheimer's in the deep learning context

Introduction

Deep learning has revolutionized CAD systems. Access to more data, better hardware, and innovation in neural networks has made deep learning methods the current state of the art in many domains. It has ushered an age of industry-grade methods that are not only accurate, but tolerant to much of the problematic data one would typically find in a CAD system.

This section unpacks deep learning within the context of AD. It starts by providing the context of the transition from traditional methods to deep learning methods, common tasks deep learning methods are used for, and what differentiates them from traditional methods. We then look at the most popular model architectures used in CAD systems and general.

The transition

Traditional CAD methods focus on feature engineering that leverage domain knowledge and maximizes discriminative power, such as features that can encapsulate medial temporal atrophy in an MRI (Frisoni et al., 2013) for classification. The methods' performance depends on how the features are engineered and whether a machine learning model is appropriate to differentiate between a positive and negative AD case. However, traditional methods can become problematic because it requires specialized knowledge and multiple stages of optimization, which may be time-consuming (Jo et al., 2019).

Deep learning-based methods take a different approach to feature engineering and classification. It uses representational learning, where features are generated "on-the-fly" during train time, thereby eliminating the need to create specialized feature preprocessing and engineering processes and achieving state-of-the-art performance (Russakovsky et al., 2015). It uses an artificial neural network (ANN), which consists of connected neurons that align the input with the subsequent output through individual weights found in each neuron and modulated through the use of an activation function, as seen in Fig. 8.2. A normal ANN (multilayer perceptron or MLP) consists of input, middle, and output layers. Still,

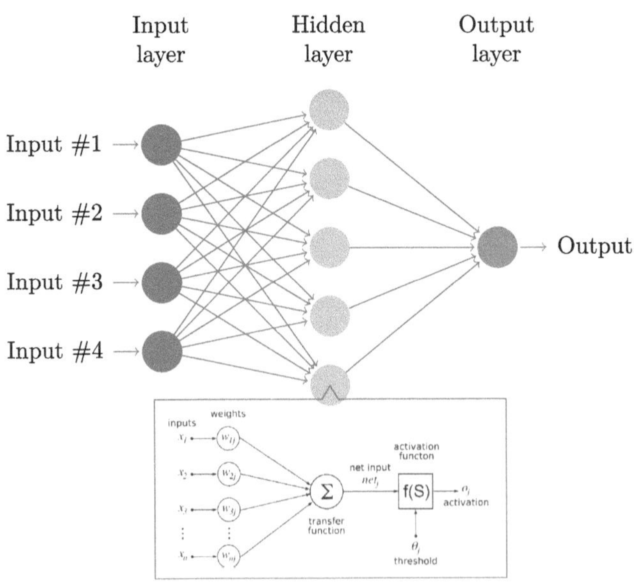

Figure 8.2 A neuron in an artificial neural network.

deep learning networks have many more hidden layers and also consist of various other parameters, types of layers, and architectures, which will be discussed in "Common model architectures" section. However, before we get to that, the next subsection discusses the tasks deep learning methods can perform in the typical CAD stages and how it compares against traditional machine learning methods.

Deep learning model tasks

Deep learning methods can be used to facilitate common tasks in the existing stages in a CAD system, such as feature extraction and classification, but it can also be used for detection or segmentation tasks, where the area of interest is isolated for further analysis. Each of these tasks is discussed in the segments below.

Segmentation

The first model task discussed is the region of interest segmentation or localization task. An area of interest is isolated or "cropped" that extracts the key area(s) found in the entire image or frame. In general, the segmentation task comes in three variants: object detection, semantic segmentation, and instance segmentation.

In object detection, the output is a rectangular bounding box containing the object of interest (with top left (x, y) coordinates, along with width and height) as shown in Fig. 8.3 and a confidence measure for objects the model is aware of (determined at train time). Traditional methods used feature templates, such as texture (De Oliveira et al., 2011), gradient difference (Abbasi & Tajeripour, 2017), or a lower-dimensional summary (Barra & Boire, 2000) on a sliding window to isolate regions of interest. Deep learning methods, such as faster regional convolutional neural networks (R-CNNs) (Ren, He, Girshick, & Sun, 2016), You only look once (YOLO) (Redmon, Divvala, Girshick, & Farhadi, 2016), and single shot detection (SSD) (Liu et al., 2016) generate proposals using a regional proposal network or equivalent and classifies the proposals using a convolutional neural network (CNN). These methods have seen recent success in isolating the right ventricle myocardium (Luo, An, Wang, Dong, & Zhang, 2016), metastatic lymph nodes, (Lu et al., 2018) and even lung nodules (George, Skaria, & Varun, 2018).

Semantic segmentation provides a class prediction for every pixel contained in the image (Xie et al., 2017), similar to the shaded area as shown

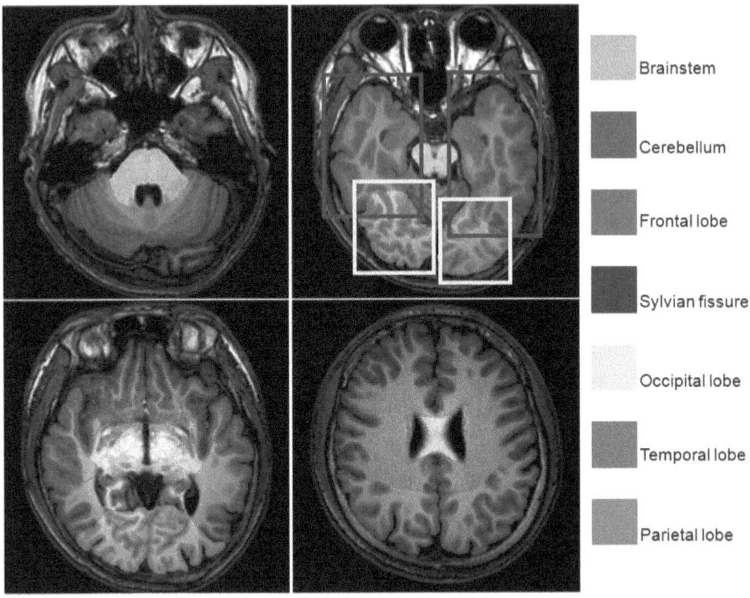

Figure 8.3 A brain MRI annotation sample depicting the regions of the brain (adapted from Zhao et al., 2015). *Taken from Zhao, H., Wang, J., Lu, Z., Wu, Q., Lv, H., Liu, H., Gong, X. (2015). Superficial siderosis of the central nervous systems induced by a single-episode of traumatic subarachnoid homerrhage: a study using mri-enhanced gradient eachi t2 star-weighted angoigraphy. PLoS One 10, e0116632.*

in Fig. 8.3, thereby providing flexibility in terms of which parts of the imagery are more important than other parts. Historically, medical technicians performed semantic segmentation manually, and it was only until the start of the century when automation of this stage became practical (Dill, Franco, & Pinho, 2015). Examples of semantic segmentation methods that use traditional methods include using atlas or template-based segmentation (Cabezas, Oliver, Lladó, Freixenet, & Cuadra, 2011) and deformable models (Cootes, Edwards, & Taylor, 2001), which have been used for the localization of the hippocampus (Crum, Scahill, & Fox, 2001), the prostrate (Toth & Madabhushi, 2012), and the medial temporal lobe (Xie et al., 2019). Deep learning methods include fully convolutional networks (Long, Shelhamer, & Darrell, 2015), UNET (Ronneberger, Fischer, & Brox, 2015), DeepLab (Chen, Papandreou, Kokkinos, Murphy, & Yuille, 2017), and hybrid CNN-CRF (conditional random field) based methods (Lin, Shen, Van Den Hengel, & Reid, 2016). Although not all of these methods have been applied in the MRI context, they have been applied for the segmentation of brain lesions (Kamnitsas, Chen, Ledig, Rueckert, & Glocker, 2015), liver tumors (Christ et al., 2017), brain tumors (Pereira, Alves, & Silva, 2018), and even certain brain regions, such as the basal ganglia and the midbrain (Milletari et al., 2017).

A more recent segmentation category, instance segmentation, aims to predict class labels and pixel-wise instance masks to localize multiple instances of objects or interesting areas (Liu, Qi, Qin, Shi, & Jia, 2018). It not only isolates the pixels belonging to a particular class, but it provides a unique identifier for each instance of the region. Current work uses methods such as Mask R-CNN (a variation of R-CNN) and methods such as YOLACT and BlendMask to achieve instance segmentation (Kim, Woo, Kim, & Kweon, 2021). Once further developed, these methods could prove invaluable in time-lapse studies with multiple objects of interest and objects that can move.

Feature extraction

Feature extraction is an unsupervised dimensional reduction method that provides a succinct representation of crucial detail in a scene while preserving attributes that make it easier for classification predictions in the subsequent stage. As mentioned before, traditional methods such as texture and gradient variants have been successfully applied for feature extraction when performing segmentation, but they can also be useful to derive features useful for classification [especially histogram of gradients (HoG) variants for AD (Sarwinda & Bustamam, 2018)]. Alternatively, other popular methods for dimensional reduction such as PCA and linear discriminant analysis, among others can be used to achieve dimensional reduction should input dimensionality or resource constraints be a problem. However, other more deep learning-based methods can also be used for dimensional reduction in the face of increased input complexity. A lower-dimensional representation is derived that minimizes reconstruction error and still preserves its inherent properties.

Examples of these methods include autoencoders and their more recent deeper variants such as variational autoencoders and adversarial autoencoders (Wang, Peng, & Ko, 2020). They obtain a compressed representation of the original input (the encoding part) and also detect repetitive structures using hidden layers that downsample every subsequent input based on priors (Wang, Yao, & Zhao, 2016). Alternatively, t-distributed stochastic neighbor embedding (t-SNE) can be used to derive a lower-dimensional representation by deriving candidate pair probability distributions, which is mapped to a lower-dimensional plane that minimizes the Kullback – Leibler divergence (Maaten & Hinton, 2008). t-SNE is mostly used to visualize high dimensional data, as shown by Gang et al. for the chest X-ray analysis of lung cancer in Gang et al. (2018). It has already been successfully applied to study the contributing parts of the brain, which relate to AD using convolutional autoencoders (Martinez-Murcia, Ortiz, Gorriz, Ramirez, & Castillo-Barnes, 2019) and getting better predictions for early prediction of AD progression (Basu et al., 2019).

Classification

The final tasks relate to the last stage in the CAD stages, classification. A decision boundary is formed by minimizing the loss function found in a supervised model using known samples and ground truth labels. As mentioned in "Classification" section there are many common traditional machine learning methods for classification that include support vector machines, random forests, and logistic regression, which only require a small number of training samples to perform predictions. Deep learning-based methods do the same but require a great deal more samples to train effective models. Common deep learning classification methods that have revolutionized AD research include CNNs (LeCun et al., 1989), which was introduced in the 1980s, but it came at significant computational cost overhead. What truly gave it momentum was when this overhead could be addressed by using the graphical processing unit (GPU) to train neural networks (Oh & Jung, 2004), and subsequently CNNs (Chellapilla, Puri, & Simard, 2006). CNN architectures such as LeNet-5 (LeCun, Bottou, Bengio, & Haffner, 1998) were finally within the reach of many more researchers, and it saw the rise of many more architectures, as shown later in "Common model architectures" section. Its application also extended beyond handwritten digits and has become a foundation in modern AD research for binary and multiclass classification.

Common model architectures

Deep neural networks have come a great deal from just using basic layers and with innovation came the introduction of new types of layers and new architectures, which saw equally impressive performance increases and further applicability of deep learning methods in newer domains. Each of the popular architectures is briefly outlined in the segments below in chronological order.

LeNet-5

As mentioned before (in "Classification" section) LeNet-5 was one of the first CNN architectures that provided significant academic contributions to the field of deep learning. Proposed by LeCun et al. in 1998 LeNet-5 used a CNN with seven layers, which consisted of three convolutional layers, two subsampling layers, and two fully connected layers, as shown in Fig. 8.4, to achieve handwritten document recognition. Convolutional layers apply a filter or kernel over parts of the respective input using a sliding window to create a feature map, encapsulating our succinct representation at varying scales. The subsampling layer serves as the dimensional reduction step typically found in a traditional pipeline. The fully connected layer transforms the input to the appropriate dimension, which relates directly to the target variable (such as having AD or not or a specific stage of progression). Another related architecture, AlexNet (Krizhevsky, Sutskever, & Hinton, 2017), was built on this by using Rectified Linear Units (ReLUs) as activation functions and dropout layers.

UNET

A very popular architecture in the biomedical community is the UNET architecture (Ronneberger et al., 2015). As mentioned in "Segmentation" section, UNET is dedicated to solving for the region of interest segmentation, where every pixel is classified according to known classes. When looking at Fig. 8.5, the U-shape gives its name. It starts with a contracting or encoding path that downsamples the input using convolution and ReLU layer blocks, followed by a max-pooling layer. The expansive or decoding path then uses

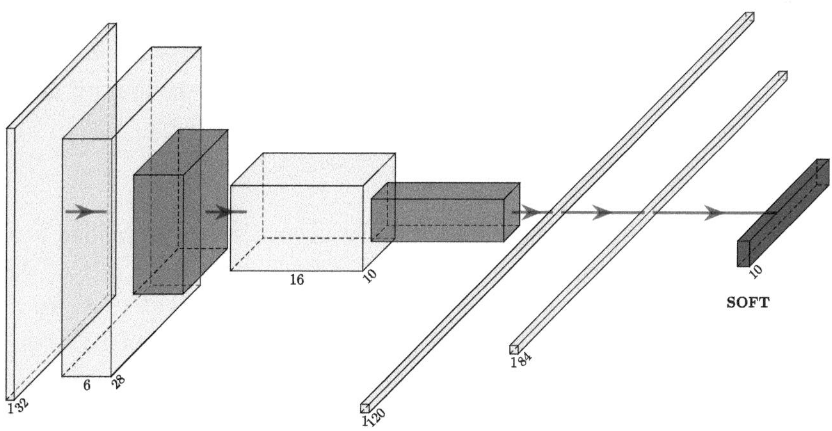

Figure 8.4 The LeNet-5 architecture. *Created using PlotNeuralNetv1 (Iqbal, H., 2018). Harisiqbal88/plotneuralnet v1.0.0, https://doi.org/10.5281/zenodo.2526396.*

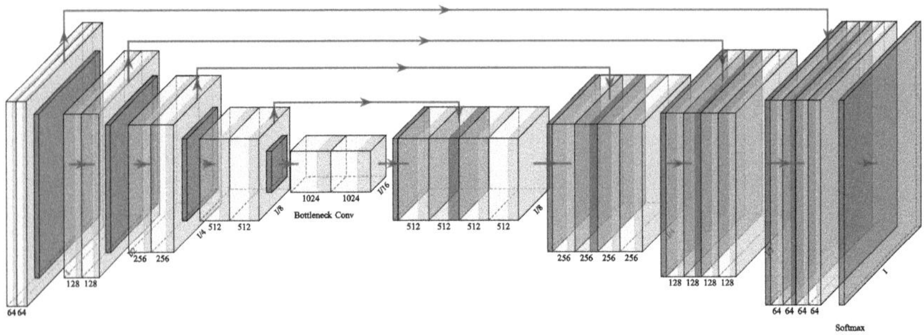

Figure 8.5 The UNET architecture. *Created using PlotNeuralNetv1 (Iqbal, H., 2018). Harisiqbal88/plotneuralnet v1.0.0, https://doi.org/10.5281/zenodo.2526396.*

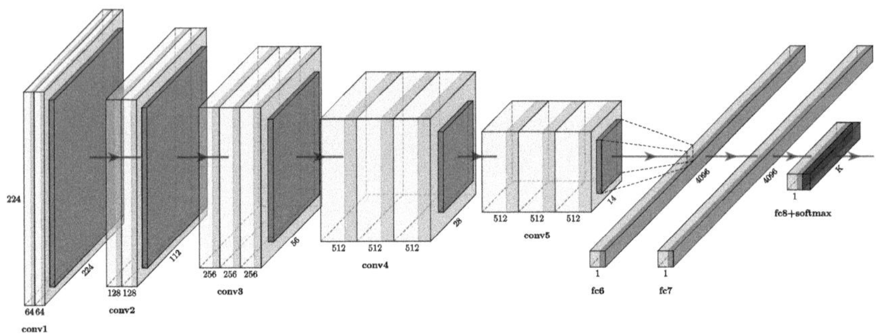

Figure 8.6 The VGG-19 architecture. *Created using PlotNeuralNetv1 (Iqbal, H., 2018). Harisiqbal88/plotneuralnet v1.0.0, https://doi.org/10.5281/zenodo.2526396.*

convolution and ReLU layer blocks with incrementing parameters and transposed convolutional blocks that upsample to the original image size to produce the output segmentation map. As Ronneberger also mentions in the original article (Ronneberger et al., 2015), UNET can also benefit from data augmentation methods, which add additional samples for training by transforming the original input in some way (rotation, rescaling, etc.).

Visual Geometry Group

Following AlexNet, researchers realized that deeper networks yielded better performance. Simonyan and Zisserman (2014), who were part of the Visual Geometry Group (VGG), took this insight and created VGG-16 in 2014, roughly twice the depth of AlexNet. It contained stacked 13 convolutional layers (with smaller filters) and three fully connected layers to get better performance, but it came at the cost of additional memory overhead. VGG-16 was followed by a deeper variant, VGG-19, as seen at the bottom of Fig. 8.6, which also had a similar effect.

ResNet

In 2015, Microsoft Research realiszd that with network depth increasing, accuracy reaches a saturation point and significantly decreases after a point, also known as the vanishing gradient problem. They solve this problem by introduction of skip connections or residuals integrated to deeper networks (as shown in Fig. 8.6), along with batch normalization, where internal layers are normalized and "recentered," which allowed for faster training and efficient gradient flow within the network (He et al., 2016).

InceptionV4

The additional resources required by VGG-16 and VGG-19 limited its scalability. Szegedy et al. (2017) solved this by packaging subnetworks using blocks or modules instead of stacking convolutional layers. By stacking blocks, which contain convolutional layers with different filters, followed by concatenation and 1×1 convolutional layers for dimensional reduction, it achieved comparable performance to VGG. The network was then further expanded using a modified optimizer, loss function, and batch normalization to specific layers, along with factorizing n by n convolutions to 1 by n and n by 1 convolutions (Szegedy, Vanhoucke, Ioffe, Shlens, & Wojna, 2016), followed by reorganization and uniformity among blocks (Szegedy et al., 2017) to give rise to Inception-v2, Inception-v3, and finally Inception-v4, shown. These were followed by variations, such as Xception (Chollet, 2017), ResNeXt-50 (Xie, Girshick, Dollár, Tu, & He, 2017) and alternatives, such as DenseNet (Huang, Liu, Van Der Maaten, & Weinberger, 2017), which serve as the current state-of-the-art architectures today.

Advantages of using deep learning

The elusive industry-grade performance in CAD systems has been a pursuit for many years, and there are quite a few reasons why many consider deep learning methods the answer to this pursuit. The subsections that follow unpack the key reasons you should use deep learning methods for AD CAD research.

End-to-end learning

The first advantage and arguably one of the most significant benefits is that there is no need for feature engineering in the pipeline. Deep learning avoids the need for hand-crafted features and instead automatically discovers the representations needed for the appropriate feature detection and classification from the raw data itself (LeCun, Bengio, & Hinton, 2015). Less time crafting features mean less development time and fewer costs.

Increased accuracy

The most attractive advantage is that deep learning methods deliver excellent results. Traditional approaches in specific fields reached a performance ceiling, and deep learning-based methods have been shown to overcome the limitations previously experienced (Guo et al., 2016). The increased performance has made deep learning a popular set of recent literature methods and integral to modern CAD systems.

Transfer learning

Transfer learning allows for the transfer of pretrained weights created for one application task and using it as a starting point in another task that does not necessarily have to have the same feature space or distribution. Transfer learning allows you to train on data you have sufficient data for and transfer the knowledge it has gained to a new network, thereby significantly improving performance on smaller datasets (Pan & Yang, 2009). The exporting of trained weights also increases the usefulness of prior trained models, thereby increasing the reusability of models and making deployment a great deal more efficient. However, there are concerns as pointed out by Guo et al. (2019).

Robustness

Missing data or adversarial attacks in CAD systems have been a long-standing problem. Guo et al. (2019) also show that deep learning methods are also reasonably robust to adversarial attacks, but results vary depending on the deep learning framework used during implementation. Through the use of data augmentation, researchers have recently shown that deep learning can also be made more robust to anomalous data input (Hendrycks et al., 2019; Gao, Saha, Prasad, & Roychoudhury, 2020), thereby further validating its use in CAD systems.

Scalability

The recent surge of demand for big data applications and the response thereof have shown that deep learning methods scale very well and can successfully find complex patterns in large amounts of data, tackle high dimensional data, semantic indexing, and other tasks in data science (Najafabadi et al., 2015). As we begin to understand the potential of deep learning methods better, so too will we unpack the additional advantages better within the field of deep learning and biomedical engineering.

Limitations of using deep learning

There are some limitations with the use of deep learning to AD. First, most deep learning models often require large datasets. This is not often available for AD. Another issue is missing data. In many medical fields, such as AD, it is not always practical to collect large amounts of data, given the frail and medical condition of many AD patients. In addition, given the old age of most AD patients, it is sometimes difficult to also collect longitudinal data from the patients, which limits the use of deep learning methods to understand disease progression.

Finally, another concern relevant to the medical community is understanding why deep learning methods result in specific predictions being performed. One strength that linear models still hold is that it is very easy to interpret why certain input translates into a respective prediction. However, the same cannot be said about more complex nonlinear models, such as deep learning-based methods. Although very recent work on methods for achieving model interpretation, such as sensitivity analysis, simple Taylor decomposition, and layer-wise relevance propagation, (Montavon, Samek, & Müller, 2018) exists, deep learning-based machine learning transparency remains a difficult task and an emerging area of exploration.

Conclusion

The age of neural networks was reborn with the rise of deep learning. The initial promise of an algorithm that mimics the visual cortex finally started paying dividends with the rise of GPU-powered training. State-of-the-art performance within CAD systems was finally within reach and the rolling out of industry commercial systems began. However, it seems computational overhead was not the only limitation of deep learning-based methods.

Although it achieved state-of-art-performance, more robustness, scalability, reusabilit,y and deployment potential, it comes at the cost of certain operational aspects in AD and CAD systems at large. The use of deep learning methods do require large datasets for adequate performance, smaller longitudinal studies can be difficult, and interpreting the model for medical understanding can be difficult. However, these limitations should not take the wind out of the AD performance sails just yet; more research is required for addressing some of our proposed recommendations and providing more value than just an answer to the clinician and medical community at large.

References

Abbasi, S., & Tajeripour, F. (2017). Detection of brain tumor in 3D MRI images using local binary patterns and histogram orientation gradient. *Neurocomputing, 219*, 526–535.

Abdi, H., & Williams, L. J. (2010). Principal component analysis. *WIREs Computational Statistics, 2*, 433–459. Available from https://doi.org/10.1002/wics0.101.

Arevalo-Rodriguez, I., Smailagic, N., Figuls, M. R. i, Ciapponi, A., Sanchez-Perez, E., Giannakou, A., ... Cullum, S. (2015). Mini-Mental State Examination (MMSE) for the detection of Alzheimer's disease and other dementias in people with mild cognitive impairment (MCI). *Cochrane Database of Systematic Reviews, 2015*(3), CD010783.

Ausó, E., Gómez-Vicente, V., & Esquiva, G. (2020). Biomarkers for Alzheimer's disease early diagnosis. *Journal of Personalized Medicine, 10*, 114.

Barra, V., & Boire, J.-Y. (2000). Tissue segmentation on MRimages of the brain by possibilistic clustering on a 3D wavelet representation. *Journal of Magnetic Resonance Imaging: An Official Journal of the International Society for Magnetic Resonance in Medicine, 11*, 267–278.

Basu S., Wagstyl, K., Zandifar, A., Collins, L., Romero, A., & Precup, D. (2019). Early prediction of Alzheimer's disease progression using variational autoencoders. In *Proceedings of the international conference on medical image computing and computer-assisted intervention*, Springer (pp. 205–213).

Bi, X., Li, S., Xiao, B., Li, Y., Wang, G., & Ma, X. (2020). Computer aided Alzheimer's disease diagnosis by an unsupervised deep learning technology. *Neurocomputing, 392*, 296–304. Available from https://doi.org/10.1016/j.neucom.2018.110.111.

Cabezas, M., Oliver, A., Lladó, X., Freixenet, J., & Cuadra, M. B. (2011). A review of atlas-based segmentation for magnetic resonance brain images. *Computer Methods and Programs in Biomedicine, 104*, e158–e177.

Chaves, R., Ramírez, J., Górriz, J. M., López, M., Salas-Gonzalez, D., Álvarez, I., & Segovia, F. (2009). SVM-based computer-aided diagnosis of the Alzheimer's disease using *t*-test NMSE feature selection with feature correlation weighting. *Neuroscience Letters, 461*, 293–297. Available from https://doi.org/10.1016/j.neulet.2009.060.052.

Chellapilla, K., Puri, S., & Simard, P. *High performance convolutional neural networks for document processing* (2006).

Chen, L.-C., Papandreou, G., Kokkinos, I., Murphy, K., & Yuille, A. L. (2017). Deeplab: Semantic image segmentation with deep convolutional nets, atrous convolution, and fully connected CRFs. *IEEE Transactions on Pattern Analysis and Machine Intelligence, 40*, 834–848.

Chollet, F. (2017). Xception: Deep learning with depthwise separable convolutions. In *Proceedings of the IEEE conference on computer vision and pattern recognition* (pp. 1251–1258).

Christ, P. F., Ettlinger, F., Grün, F., Elshaera, M. E. A., Lipkova, J., Schlecht, S. ... Bilic, P. (2017). Automatic liver and tumor segmentation of CT and MRI volumes using cascaded fully convolutional neural networks, arXiv: 1702.05970.

Cootes, T. F., Edwards, G. J., & Taylor, C. J. (2001). Active appearance models. *IEEE Transactions on Pattern Analysis and Machine Intelligence, 23*, 681–685.

Crum, W. R., Scahill, R. I., & Fox, N. C. (2001). Automated hippocampal segmentation by regional fluid registration of serial MRI: Validation and application in Alzheimer's disease. *NeuroImage, 13*, 847–855.

De Oliveira, M., Balthazar, M., D'abreu, A., Yasuda, C., Damasceno, B., Cendes, F., & Castellano, G. (2011). MR imaging texture analysis of the corpus callosum and thalamus in amnestic mild cognitive impairment and mild Alzheimer disease. *American Journal of Neuroradiology, 32*, 60–66.

Dessouky, M. M., Elrashidy, M. A., Taha, T. E., & Abdelkader, H. M. (2016). Computer-aided diagnosis system for Alzheimer's disease using different discrete transform techniques. *American Journal of Alzheimer's Disease & Other Dementiasr, 31*, 282–293, SAGE Publications Inc. Available from https://doi.org/10.1177/1533317515603957.

Dill, V., Franco, A. R., & Pinho, M. S. (2015). Automated methods for hippocampus segmentation: The evolution and a review of the state of the art. *Neuroinformatics, 13*, 133–150.

Ding, Y., Sohn, J. H., Kawczynski, M. G., Trivedi, H., Harnish, R., Jenkins, N. W., ... Franc, B. L. (2018). A deep learning model to predict a diagnosis of Alzheimer disease by using 18F-FDG PET of the brain. *Radiology, 290*, 456–464, Radiological Society of North America. Available from https://doi.org/10.1148/radiol.2018180958.

Dubois, B., Feldman, H. H., Jacova, C., DeKosky, S. T., Barberger-Gateau, P., Cummings, J., ... Jicha, G. (2007). Research criteria for the diagnosis of Alzheimer's disease: Revising the NINCDS-ADRDA criteria. *The Lancet Neurology, 6*, 734–746.

Ebrahimighahnavieh, M. A., Luo, S., & Chiong, R. (2020). Deep learning to detect Alzheimer's disease from neuroimaging: A systematic literature review. *Computer Methods and Programs in Biomedicine, 187*, 105242. Available from https://doi.org/10.1016/j.cmpb.2019.105242.

Frisoni, G. B., Bocchetta, M., Chételat, G., Rabinovici, G. D., De Leon, M. J., Kaye, J., ... Black, S. E. (2013). Imaging markers for Alzheimer disease: Which vs how. *Neurology, 81*, 487–500.

Gang, P. Zhen, W., Zeng, W., Gordienko, Y., Kochura, Y., Alienin, O. ... Stirenko, S. (2018). Dimensionality reduction in deep learning for chest X-ray analysis of lung cancer. In *Proceedings of the 2018 tenth international conference on advanced computational intelligence (ICACI)*, IEEE (pp. 878–883).

Gao, X., Saha R. K., Prasad, M. R., & Roychoudhury, A. (2020). Fuzz testing based data augmentation to improve robustness of deep neural networks. In *Proceedings of the 2020 IEEE/ACM 42nd international conference on software engineering (ICSE)*, IEEE (pp. 1147–1158).

George, J., Skaria, S., & Varun, V. (2018). Using YOLO based deep learning network for real time detection and localization of lung nodules from low dose CT scans. In *Medical imaging 2018: computer-aided diagnosis*, proceedings volume 10575, International Society for Optics and Photonics (p. 105751I).

Guo, Q., Chen, S., Xie, X., Ma, L., Hu, Q., Liu, H. ... Li, X. (2019). An empirical study towards characterizing deep learning development and deployment across different frameworks and platforms. In *Proceedings of the 2019 34th IEEE/ACM international conference on automated software engineering (ASE)*, IEEE (pp. 810–822).
Guo, Y., Liu, Y., Oerlemans, A., Lao, S., Wu, S., & Lew, M. S. (2016). Deep learning for visual understanding: A review. *Neurocomputing, 187*, 27–48.
He, K., Zhang, X., Ren, S., & Sun, J. (2016). Deep residual learning for image recognition. In *Proceedings of the IEEE conference on computer vision and pattern recognition* (pp. 770–778).
Hendrycks, D., Mu, N., Cubuk, E. D., Zoph, B., Gilmer, J., & Lakshminarayanan, B. (2019). *Augmix: A simple data processing method to improve robustness and uncertainty*, arXiv:1912.02781.
Huang, G., Liu, Z., Van Der Maaten, L., & Weinberger, K. Q. (2017). Densely connected convolutional networks. In: *Proceedings of the IEEE conference on computer vision and pattern recognition* (pp. 4700–4708).
Hulbert, S., & Adeli, H. (2013). EEG/MEG- and imaging-based diagnosis of Alzheimer's disease. *Reviews in the Neurosciences, 24*, 563–576, De Gruyter section: Reviews in the Neurosciences. Available from https://doi.org/10.1515/revneuro-2013-0042.
Iqbal, H. (2018). Harisiqbal88/plotneuralnet v1.0.0. Available from https://doi.org/10.5281/zenodo.2526396.
James, B. D., Power, M. C., Gianattasio, K. Z., Lamar, M., Oveisgharan, S., Shah, R. C., ... Bennett, D. A. (2020). Characterizing clinical misdiagnosis of dementia using medicare claims records linked to Rush Alzheimer's Disease Center (radc) cohort study data: Public health: Innovative methods in ADRD research. *Alzheimer's & Dementia, 16*, e044880.
Jo, T., Nho, K., & Saykin, A. J. (2019). Deep learning in Alzheimer's disease: Diagnostic classification and prognostic prediction using neuroimaging data. *Frontiers in Aging Neuroscience, 11*, 220.
Kamnitsas, K., Chen, L., Ledig, C., Rueckert, D., & Glocker, B. (2015). Multi-scale 3D convolutional neural networks for lesion segmentation in brain MRI. *Ischemic Stroke Lesion Segmentation, 13*, 46.
Karami, V., Nittari, G., & Amenta, F. (2019). Neuroimaging computer-aided diagnosis systems for Alzheimer's disease. *International Journal of Imaging Systems and Technology, 29*, 83–94. Available from https://doi.org/10.1002/ima.22300.
Khalid, S., Khalil, T., & Nasreen, S. (2014). A survey of feature selection and feature extraction techniques in machine learning. In *Proceedings of the 2014 science and information conference (SAI)*, IEEE, London, UK (pp. 372–378). <https://ieeexplore.ieee.org/document/6918213> (Online; accessed 11.12.20).
Kim, M., Woo, S., Kim, D., & Kweon, I. S. (2021). The devil is in the boundary: Exploiting boundary representation for basis-based instance segmentation. In *Proceedings of the IEEE/CVF winter conference on applications of computer vision* (pp. 929–938).
Krizhevsky, A., Sutskever, I., & Hinton, G. E. (2017). Imagenet classification with deep convolutional neural networks. *Communications of the ACM, 60*, 84–90.
Krupinski, E. A. (2020). Evaluating AI clinically—it's not just ROC AUC! *Radiology* (p. 203782), Radiological Society of North America. Available from https://doi.org/10.1148/radiol.2020203782.
Lahmiri, S., & Shmuel, A. (2019). Performance of machine learning methods applied to structural MRI and ADASs cognitive scores in diagnosing Alzheimer's disease. *Biomedical Signal Processing and Control, 52*, 414–419. Available from https://doi.org/10.1016/j.bspc.2018.080.009.
LeCun, Y., Bengio, Y., & Hinton, G. (2015). Deep learning. *Nature, 521*, 436–444.

LeCun, Y., Boser, B., Denker, J. S., Henderson, D., Howard, R. E., Hubbard, W., & Jackel, L. D. (1989). Backpropagation applied to handwritten zip code recognition. *Neural Computation, 1*, 541–551.
LeCun, Y., Bottou, L., Bengio, Y., & Haffner, P. (1998). Gradient-based learning applied to document recognition. *Proceedings of the IEEE, 86*, 2278–2324.
Lin, G., Shen, C., Van Den Hengel, A., & Reid, I. (2016). Efficient piecewise training of deep structured models for semantic segmentation. In *Proceedings of the IEEE conference on computer vision and pattern recognition* (pp. 3194–3203).
Liu, S., Qi, L., Qin, H., Shi, J., & Jia, J. (2018). Path aggregation network for instance segmentation. In *Proceedings of the IEEE conference on computer vision and pattern recognition* (pp. 8759–8768).
Liu, W., Anguelov, D., Erhan, D., Szegedy, C., Reed, S., Fu, C.-Y., & Berg, A. C. (2016). SSD: Single shot multibox detector. In *Proceedings of the European conference on computer vision*, Springer (pp. 21–37).
Long, J., Shelhamer, E., & Darrell, T. (2015). Fully convolutional networks for semantic segmentation. In *Proceedings of the IEEE conference on computer vision and pattern recognition* (pp. 3431–3440).
Lu, Y., Yu, Q., Gao, Y., Zhou, Y., Liu, G., Dong, Q., ... Zhang, Z. (2018). Identification of metastatic lymph nodes in MR imaging with faster region-based convolutional neural networks. *Cancer Research, 78*, 5135–5143.
Luo, G., An, R., Wang, K., Dong, S., & Zhang, H. (2016). A deep learning network for right ventricle segmentation in short-axis MRI. In *Proceedings of the 2016 computing in cardiology conference (CinC)*, IEEE (pp. 485–488).
Maaten, L. v d, & Hinton, G. (2008). Visualizing data using t-sne. *Journal of Machine Learning Research, 9*, 2579–2605.
Martinez-Murcia, F. J., Ortiz, A., Gorriz, J.-M., Ramirez, J., & Castillo-Barnes, D. (2019). Studying the manifold structure of Alzheimer's disease: A deep learning approach using convolutional autoencoders. *IEEE Journal of Biomedical and Health Informatics, 24*, 17–26.
Milletari, F., Ahmadi, S.-A., Kroll, C., Plate, A., Rozanski, V., Maiostre, J., ... Bötzel, K. (2017). Hough-CNN: Deep learning for segmentation of deep brain regions in MRI and ultrasound. *Computer Vision and Image Understanding, 164*, 92–102.
Miotto, R., Wang, F., Wang, S., Jiang, X., & Dudley, J. T. (2018). Deep learning for healthcare: Review, opportunities and challenges. *Briefings in Bioinformatics, 19*, 1236–1246, publisher: Oxford Academic. Available from https://doi.org/10.1093/bib/bbx044.
Montavon, G., Samek, W., & Müller, K.-R. (2018). Methods for interpreting and understanding deep neural networks. *Digital Signal Processing, 73*, 1–15.
Najafabadi, M. M., Villanustre, F., Khoshgoftaar, T. M., Seliya, N., Wald, R., & Muharemagic, E. (2015). Deep learning applications and challenges in big data analytics. *Journal of Big Data, 2*, 1–21.
Oh, K.-S., & Jung, K. (2004). GPU implementation of neural networks. *Pattern Recognition, 37*, 1311–1314.
Pan, S. J., & Yang, Q. (2009). A survey on transfer learning. *IEEE Transactions on Knowledge and Data Engineering, 22*, 1345–1359.
Pellegrini, E., Ballerini, L., Hernandez, Md. C. V., Chappell, F. M., Gonzáilez-Castro, V., Anblagan, D., ... Wardlaw, J. M. (2018). Machine learning of neuroimaging for assisted diagnosis of cognitive impairment and dementia. *A systematic review, Alzheimer's & Dementia: Diagnosis, Assessment & Disease Monitoring, 10*, 519–535. Available from https://doi.org/10.1016/j.dadm.2018.070.004.
Pereira, S., Alves, V., & Silva, C.A. (2018). Adaptive feature recombination and recalibration for semantic segmentation: Application to brain tumor segmentation in MRI. In *Proceedings of international conference on medical image computing and computer-assisted intervention*, Springer (pp. 706–714).

Podgorelec, V. (2012). Analyzing EEG signals with machine learning for diagnosing Alzheimer's disease. *Elektronika ir Elektrotechnika, 18*(8), 61–64. Available from https://doi.org/10.5755/j01.eee.18.8.2627.

Raghavendra, U., Acharya, U. R., & Adeli, H. (2019). Artificial intelligence techniques for automated diagnosis of neurological disorders. *European Neurology, 82*, 41–64, publisher: Karger Publishers PMID: 31743905. Available from https://doi.org/10.1159/000504292.

Redmon, J., Divvala, S., Girshick, R., & Farhadi, A. (2016). You only look once: Unified, real-time object detection. In *Proceedings of the IEEE conference on computer vision and ppattern recognition* (pp. 779–788).

Ren, S., He, K., Girshick, R., & Sun, J. (2016). Faster R-CNN: Towards real-time object detection with region proposal networks. *IEEE Transactions on Pattern Analysis and Machine Intelligence, 39*, 1137–1149.

Ronneberger, O., Fischer, P., & Brox, T. (2015). U-net: Convolutional networks for biomedical image segmentation. In *Proceedings of the international conference on medical image computing and computer-assisted intervention*, Springer (pp. 234–241).

Ruiz, E., Ramírez, J., Górriz, J. M., Casillas, J., & Initiative, tA. D. N. (2018). Alzheimer's disease computer-aided diagnosis: Histogram-based analysis of regional mri volumes for feature selection and classification. *Journal of Alzheimer's Disease, 65*, 819–842, publisher: IOS Press. Available from https://doi.org/10.3233/JAD-170514.

Russakovsky, O., Deng, J., Su, H., Krause, J., Satheesh, S., Ma, S., ... Bernstein, M. (2015). Imagenet large scale visual recognition challenge. *International Journal of Computer Vision, 115*, 211–252.

Sarwinda, D., & Bustamam A. (2018) 3D-HoG features-based classification using MRI images to early diagnosis of Alzheimer's disease. In *Proceedings of the 2018 IEEE/ACIS 17th international conference on computer and information science* (ICIS), IEEE (pp. 457–462).

Segovia, F., Bastin, C., Salmon, E., Górriz, J. M., Ramírez, J., & Phillips, C. (2014). Combining PET images and neuropsychological test data for automatic diagnosis of Alzheimer's disease. *PLoS One, 9*, e88687, publisher: Public Library of Science. Available from https://doi.org/10.1371/journal.pone.0088687.

Shaikh, T. A., & Ali, R. (2019). Automated atrophy assessment for Alzheimer's disease diagnosis from brain MRI images. *Magnetic Resonance Imaging, 62*, 167–173. Available from https://doi.org/10.1016/j.mri.2019.060.019.

Shakarami, A., Tarrah, H., & Mahdavi-Hormat, A. (2020). A CAD system for diagnosing Alzheimer's disease using 2D slices and an improved AlexNet-SVM method. *Optik, 212*, 164237. Available from https://doi.org/10.1016/j.ijleo.2020.164237.

Shatte, A. B. R., Hutchinson, D. M., & Teague, S. J. (2019). Machine learning in mental health: A scoping review of methods and applications. *Psychological Medicine 49*, 1426–1448, publisher: Cambridge University Press. Available from https://doi.org/10.1017/S0033291719000151.

Simonyan, K., & Zisserman, A. (2014). *Very deep convolutional networks for large-scale image recognition*, arXiv:1409.1556.

Subasi, A. (2020). Use of artificial intelligence in Alzheimer's disease detection. *In Artificial intelligence in precision health* (pp. 257–278).

Szegedy, C., Ioffe, S., Vanhoucke, V., & Alemi, A. (2017). Inception-v4, inception-ResNet and the impact of residual connections on learning. In *Proceedings of the AAAI conference on artificial intelligence*, Vol. 31.

Szegedy, C., Vanhoucke, V., Ioffe, S., Shlens, J., & Wojna, Z. (2016). Rethinking the inception architecture for computer vision. In *Proceedings of the IEEE conference on computer vision and pattern recognition* (pp. 2818–2826).

Tanveer, M., Richhariya, B., Khan, R. U., Rashid, A. H., Khanna, P., Prasad, M., & Lin, C. T. (2020). Machine learning techniques for the diagnosis of Alzheimer's disease: A review. *ACM Transactions on Multimedia Computing, Communications, and Applications, 16*, 30:1–30:35. Available from https://doi.org/10.1145/3344998.

Toth, R., & Madabhushi, A. (2012). Multifeature landmark-free active appearance models: Application to prostate MRI segmentation. *IEEE Transactions on Medical Imaging, 31*, 1638–1650.

Valenzuela, O., Jiang, X., Carrillo, A., & Rojas, I. (2018). Multi-objective genetic algorithms to find most relevant volumes of the brain related to Alzheimer's disease and mild cognitive impairment. *International Journal of Neural Systems, 28*, 1850022, publisher: World Scientific Publishing Co. Available from https://doi.org/10.1142/S0129065718500223.

Wang, H., & Raj, B. (2017). *On the origin of deep learning*, arXiv:1702.07800.

Wang, H.-P., Peng, W.-H., & Ko, W.-J. (2020). Learning priors for adversarial autoencoders. *APSIPA Transactions on Signal and Information Processing, 9*.

Wang, Y., Yao, H., & Zhao, S. (2016). Auto-encoder based dimensionality reduction. *Neurocomputing, 184*, 232–242.

WHO. (2019). *Cause-specific mortality, 2000–2019*. https://www.who.int/data/gho/data/themes/mortality-and-global-health-estimates/ghe-leading-causes-of-death [Online]. Accessed 15.05.21.

Xie, C., Wang, J., Zhang, Z., Zhou Y., Xie, L., & Yuille, A. (2017). Adversarial examples for semantic segmentation and object detection. In *Proceedings of the IEEE international conference on computer vision* (pp. 1369–1378).

Xie, L., Wisse, L. E., Pluta, J., de Flores, R., Piskin, V., Manjón, J. V., ... Wolk, D. A. (2019). Automated segmentation of medial temporal lobe subregions on in vivo T1-weighted MRI in early stages of Alzheimer's disease. *Human Brain Mapping, 40*, 3431–3451.

Xie, S., Girshick, R., Dollár, P., Tu, Z., & He, K. (2017). Aggregated residual transformations for deep neural networks. In *Proceedings of the IEEE conference on computer vision and pattern recognition* (pp. 1492–1500).

Xu, Y., Chen, K., Zhao, Q., Li, F., & Guo, Q. (2018). Short-term delayed recall of auditory verbal learning test provides equivalent value to long-term delayed recall in predicting MCI clinical outcomes: A longitudinal follow-up study. *Applied Neuropsychology: Adult, 27*(1), 73–81.

Zhang, F., Tian, S., Chen, S., Ma, Y., Li, X., & Guo, X. (2019). Voxel-based morphometry: Improving the diagnosis of Alzheimer's disease based on an extreme learning machine method from the ADNI cohort. *Neuroscience, 414*, 273–279. Available from https://doi.org/10.1016/j.neuroscience.2019.050.014.

PART III

Treatment of dementia

CHAPTER 9

Treatment of depression in Alzheimer's disease

Ahmed A. Moustafa[1,2,3], Lily Bilson[1] and Wafa Jaroudi[4]
[1]School of Psychology, Western Sydney University, Sydney, NSW, Australia
[2]MARCS Institute for Brain and Behaviour, Western Sydney University, Sydney, NSW, Australia
[3]Department of Human Anatomy and Physiology, The Faculty of Health Sciences, University of Johannesburg, Johannesburg, South Africa
[4]School of Social Sciences and Psychology, Western Sydney University, Sydney, NSW, Australia

Introduction

With the estimated cost of US$172 billion per year for patients with Alzheimer's disease (AD), finding successful treatments that reduce the burden of caring for AD patients is of the utmost importance to society as well as the patients themselves and their families.

Several studies have also shown that depression can be a risk factor for the development of several neurological disorders, including AD (Bennett & Thomas, 2014; Linnemann & Lang, 2020; Modrego & Ferrandez, 2004; Steck, Cooper, & Orgeta, 2018).

The prevalence rates of depression in AD vary among several studies. Barca and colleagues (Barca, Selbaek, Laks, & Engedal, 2008) report that the prevalence rate of depression in dementia is as high as 86%. However, in other studies, depressive symptoms afflict between 17% (Cummings, Miller, Hill, & Neshkes, 1987) and 87% (Merriam, Aronson, Gaston, Wey, & Katz, 1988) of patients with AD. However, in other studies, it was reported that depression prevalence estimates in AD vary between 26% and 50% (Starkstein et al., 1997; Teng et al., 2008) with the best indicators of depression in dementia being persistent sad mood or decreased positive affect (Teng et al., 2008). Given its high rate, it is important to provide a suitable treatment to manage depression in AD, which would help decrease memory decline in the patients, as argued by several studies, which we discuss in the following.

Treatment

In this section we discuss the several existing treatment strategies for managing depression in AD. There have been several studies employing

Alzheimer's Disease
DOI: https://doi.org/10.1016/B978-0-12-821334-6.00012-0

different treatment methods to ameliorate depressive symptoms in dementia patients (for a review see Orgeta, Qazi, Spector, & Orrell, 2015). For a discussion on several pharmacological and nonpharmacological treatments options for depression in AD see Gellis, McClive-Reed, and Brown (2009). In the following we discuss both pharmacological and behavioral treatments to manage depressive symptoms in AD.

Pharmacological treatment

There are a multitude of studies on the use of several drugs to treat depression in dementia (Huang et al., 2020), including antidepressants, anxiolytics, and anticholinergic medications. Some of these are either administered by themselves or in combination with antidepressants. For example, anxiolytics and hypnotics were often used in lieu of, or in addition to antidepressant therapy (Evers et al., 2002). Some studies have shown that anticholinergic medications can also lessen depressive symptoms in dementia patients (Birks, 2006; Mowla, Mosavinasab, Haghshenas, & Borhani Haghighi, 2007).

The use of antidepressants to treat depression in dementia is controversial as there is insufficient evidence regarding their efficacy (Nagata, Shinagawa, Nakajima, Noda, & Mimura, 2020). However, selective serotonin reuptake inhibitors (SSRIs) are frequently prescribed and tend to be well tolerated. A close monitoring of the efficacy and side effects of SSRIs is recommended. Several studies have shown that many antidepressants (e.g., sertraline, fluoxetine, citalopram, trazodone, moclobemide, citalopram, trazadone, buspirone, and mirtazapine and sertraline) can ameliorate depressive symptoms in dementia (Caballero, Hitchcock, Beversdorf, Scharre, & Nahata, 2006; Herrmann & Lanctot, 2007; Lyketsos et al., 2003; Menon et al., 2001; Sink, Holden, & Yaffe, 2005; Starkstein & Mizrahi, 2006), but for different results see Evers et al. (2002). Older studies have also found that tricyclic antidepressants can ameliorate depressive symptoms in AD patients (Reifler et al., 1989; Teri et al., 1991). For example, Menon et al. (2001) recommended the use of SSRI antidepressants, arguing that they would be the most efficacious as depression is related to dysregulation of the serotoninergic system. A recent study compared the effectiveness of several antidepressants and vascular medicines on managing depression in patients with dementia, including sertraline, escitalopram, and nicergoline (Takemoto et al., 2020). The study found escitalopram had a better effect in reducing depression than other drugs. These results are somewhat in agreement with Gellis et al. (2009), which reported that citalopram and sertraline are possibly more effective at treating depressive symptoms in dementia than other antidepressants.

Unlike prior studies, according to Evers et al. (2002), 42% of dementia patients who received a diagnosis of a depressive disorder were not treated with antidepressants. Evers et al. (2002) also discussed some reports on the common overuse and misuse of these drugs for the treatment of depression in the elderly (Steffens et al., 2000). Furthermore, Nelson and Devanand (2011) found that only antidepressants has a minor effect on decreasing depression in dementia patients; especially they conducted a meta-analysis and found that only two out of seven studies reported that antidepressants are better than placebo in treating depression (for discussion see Lenze, 2011). It has been suggested that antidepressants can manage dementia and memory decline as they also ameliorate hippocampus function (Dafsari & Jessen, 2020).

In sum, there are several pharmacological treatments for depression in AD, including SSRIs as well as anxiolytics. However, findings are not conclusive regarding the efficacy of these medications for managing depression in AD. In the next section, we will discuss existing behavioral treatment for managing depression in AD.

Behavioral treatment

There are various behavioral treatments for depression in dementia, including exercises, cognitive training, reminiscence therapy, and cognitive behavioral therapy (CBT).

Williams and colleagues (Williams & Tappen, 2007, 2008) have investigated the effects of a 16-week comprehensive exercise program on depressive symptoms in a sample of 45 nursing home residents with moderate to severe AD. The study had a three-group: Group 1 individuals have conducted comprehensive exercise, Group 2 individuals have conducted supervised walking, and Group 3 individuals have conducted social conversation. Participants were required to be able to walk with assistance. Those who could walk unassisted were excluded. Depression was measured by the Cornell Scale for Depression in Dementia (CSDD). Participants were required to score 7 or above to be eligible to participate in the study. Mood, defined as prolonged emotional state, was measured using the Dementia Mood Assessment Scale and the Alzheimer's Mood Scale. Affect, defined as the external expression of emotions, was measure by the Observed Affect Scale, which has been previously used with AD patients (Lawton, Van Haitsma, & Klapper, 1996). Group 1 underwent two programs.

First, it included 10 minutes of strength, balance, and flexibility training, increasing the number of repetitions each week. Second, there was a walking component, which was increased in duration each week. Group 2 were given individual walking paces according to ability, with sessions increasing in duration each week until a maximum of 30 minutes. Group 3 was the attention control group, engaging in casual conversation for an equivalent amount of time. The results found that the mean score on the CSDD decreased from 12.40 at baseline to a mean of 9.77 across all the three treatment groups. Thirty-five percent of the participants had CSDD scores under 7 at posttest. There were significant reductions in scores across all but one of the outcome measures from baseline to posttest in the participant sample as whole. The control group evidently had some beneficial effect. Exercise programs aimed at treating depression in nursing home residents with severe AD have proven to have a clear benefit. These results are also in agreement with other studies reporting benefit of exercise in treating depression in dementia (Regan, Katona, Walker, & Livingston, 2005). Specifically, Regan et al. (2005) found that dementia patients who participate in exercise activities tend to report less depressive symptoms than those who do not do exercises. Due to the nature of exercise being difficult to implement with those who are frequently unwell, the benefits of chair-based exercise should be investigated as an alternative to active exercise.

Stewart et al. (2017) have investigated the efficacy of a nonpharmacological, psychosocial group intervention known as cognitive stimulation therapy (CST) in improving cognitive function, quality of life, and well-being in indivuals with dementia. The authors assessed the treatment's impact on cognition, quality of life, and depression in six groups, totaling 40 individuals. CST incorporates reminiscence therapy, reality orientation, and validation therapy in a structured psychosocial program whereby their opinions are valued and they are cognitively challenged. CST comprises 14 themed sessions and activities (approximately 45 minutes long), administered twice a week for 7 weeks. Cognitive tasks involve executive function, multisensory stimulation, and reminiscence as an aid for orientation. Activity themes include physical games, sounds, childhood memories, food, current affairs, faces and scenes, word association, creativity, categorizing objects, orientation, using money, number games, word games, and team games. The key principles that make up the foundation of the therapy include mental stimulation, new ideas and thoughts, using orientation sensitively and implicitly, using reminiscence as an aid to the here-and-now, providing triggers for recall, continuity, and consistency between sessions, implicit learning, stimulating language, person-centered, respect, involvement, inclusion, choice, fun, maximizing potential, and building relationships. Assessments were completed at two points: once before the 7-week intervention and once at the end of the 7-week intervention. The age range of the participants ranges from 50 to 94 years old (mean = 78.8). Depression was assessed using the CSDD using direct observation and interview. Clinicians noted various mood-related signs, behavioral disturbances, physical symptoms, cyclic functions, and ideational disturbances. There was a significant difference in the Cornell Scale score pre- and postintervention. Participants exhibited lessened depression and improved cognitive status and were observably more talkative, expressed more joy and developed friendships with their peers. Overall, CST is an effective means of alleviating depression in dementia. Reminiscence therapy provides dementia patients with aids to remember past events. If successful to remember them, it is very likely that enhancement of memory will also be associated with reduction in depressive symptoms. These predictions regarding a positive effect of reminiscence therapy in dementia have been confirmed in a recent study (Moon & Park, 2020). However, a recent review has revealed some doubt on the effect of reminiscence therapy on reducing depression in AD. Woods, O'Philbin, Farrell, Spector, and Orrell (2018) reported inconsistent results among 22 studies conducted on the impact of reminiscence therapy on dementia.

Marriott, Donaldson, Tarrier, and Burns (2000) assessed whether family intervention (described in the following) reduces the subjective burden of care in carers of patients with AD and whether, this in turn, produces clinical benefits in the patients themselves. The primary hypothesis of the study was that family intervention will reduce the burden of care and secondarily that the intervention would show an effect on cognitive and noncognitive symptoms in the patients, including depression. A randomized control trial of family intervention was compared with two control groups with blind and independent assessment. The intervention consisted of three components—carer education, stress management, and coping skills training—and was spread over 14 sessions with 2-week intervals between each session. The Camberwell Family Interview (CFI) was conducted on participants in two of the three groups, one experimental and one control. The interview was audiotaped and semistructured from which the rating of expressed emotion was derived and took approximately 90 minutes. The following assessments were carried out at pretreatment, posttreatment and 3-month follow-up on the carers: General Health Questionnaire and Beck Depression Inventory. At the same intervals on the patients the following assessments were carried out: Mini-Mental State Examination, CSDD, MOUSEPAD (assesses a number of psychiatric symptoms and behavioral disturbances in dementia), and the Clinical Dementia Rating. The primary hypothesis was confirmed in that the carers in the intervention group recorded significant improvements on the measures of distress and depression at posttreatment and 3-month follow-up when compared with the two control groups. Findings of the study indicated that there was no effect of the CFI-only control group indicating that an one-off session is not sufficient to reduce carer burden. There was a modest support for the secondary hypothesis in that there was a significant effect of family intervention at posttreatment on behavioral disturbance; however, this was not maintained at follow-up. Both behavioral interventions significantly reduced depression in patients and carers. The clinical implication is that interventions that help carers cope better with their own emotional states and aid a clearer perspective on ways in which to deal with challenging situations are empirically beneficial. An improved feeling of being able to cope and a sense of control in the situation brings about positive consequences for the carer and the patient.

Although CBT is commonly used with patients with major depressive disorders (Ahern, Kinsella, & Semkovska, 2018; Rubin-Falcone et al., 2018; Vara et al., 2018; Zhang, Zhang, Zhang, Jin, & Zheng, 2018), it is less commonly used to treat depressive symptoms in dementia. Teri and colleagues have found evidence that CBT can indeed reduce depressive symptoms in dementia patients (Teri & Gallagher-Thompson, 1991; Teri, Logsdon, Uomoto, & McCurry, 1997). There are few more recent studies as well. Walker (2004) has also reported a case study of a patient who benefited greatly from CBT that has included altering beliefs associated with dementia and cognitive decline. More recently, Forstmeier, Maercker, Savaskan, and Roth (2015) suggested that some elements of CBT including engagement in pleasant activities and life review (i.e., going through different stages of life events to enhance memorizing) would help lessen depressive symptoms in dementia patients. It is also important to note there are several studies on the efficacy of CBT to treat depressive symptoms in the caregivers with dementia patients (for a review see, Hopkinson, Reavell, Lane, & Mallikarjun, 2018, although this is beyond the scope of the current review). Along these lines, one recent study found that a mindfulness-based cognitive therapy is better than a mindfulness-based stress reduction therapy in reducing depression in patients with dementia (Cheung et al., 2020).

Another study have used cognitive training to ameliorate cognitive decline and depression in individuals with mild cognitive impairment (Sukontapol et al., 2018). Sukontapol and colleagues found that unlike control treatment, cognitive training reduces depressive symptoms. In another study, Smith-Ray, Irmiter, and Boulter (2016) found that 10 weeks of cognitive training (using Posit Science Brain Health Questionnaire for 2 hours a day) led to a reduction in depression in older adults at risk for developing dementia. For a review on the impact of cognitive training in older adults as well as individuals at risk to develop dementia, see Naismith and Mowszowski (2016). For a review on how cognitive training can manage depression in dementia, see Chan et al. (2020).

Few studies have also used the Lewinsohn's pleasant events training method, which attempt to explain to the patients that their behavior is related to how they feel and help the patients decrease negative feelings. What is key here is that training method can be conducted by the caregivers at home or nursing homes. One study has reported that the Lewinsohn's pleasant events training method is quite effective in reducing depressive symptoms in dementia patients (Teri et al., 2003). However, another study did not report any positive results with the use of Lewinsohn's pleasant events training method (Lichtenberg, Kemp-Havican, Macneill, & Schafer Johnson, 2005).

Some other studies have utilized other kinds of therapies. In one study, it was reported that bright light therapy can effectively reduce depressive symptoms especially in patients with severe dementia (Onega, Pierce, & Epperly, 2018). There have been several studies on how pets can help manage depression in young and older adults (Ambrosi, Zaiontz, Peragine, Sarchi, & Bona, 2019; Mota Pereira & Fonte, 2018). Along these lines, animal-assisted therapy was found to manage depression in dementia patients (Baek, Lee, & Sohng, 2020).

Some other studies have investigated the impact of combined treatments (one behavioral and one pharmacological) on reducing depression in dementia. A recent study has investigated the impact of both antidepressants and social engagement on depression in dementia patients (Rodrigues, Capuano, Barnes, Bennett, & Shah, 2018). Social engagement measure included both social network size (i.e., number of friends, family a person has, and how often they see them) and perceived social isolation (i.e., loneliness). Rodrigues et al. (2018) found that social engagement, but not antidepressants, decreased depressive symptoms. In another recent study, it was reported that adding active music therapy to memantine is more effective than memantine alone in reducing depression in dementia (Giovagnoli et al., 2018). A recent review of 82 studies concluded that music therapy can treat depression and anxiety in dementia patients (Lam, Li, Laher, & Wong, 2020). Similar findings were reported in another review (Moreno-Morales, Calero, Moreno-Morales, & Pintado, 2020). In a recent study, it was found that a behavioral treatment that includes cognitive training, art therapy, and music therapy can effectively manage depression in dementia patients (Jung et al., 2020).

In sum, there are several nonpharmacological treatments for depression in AD, which include, but not limited to, exercise therapy, CBT, family-based interventions, among others. Some, but not all, are effective at managing depression in AD.

Conclusions

In sum, our literature review shows that many of the existing behavioral therapies are much better than pharmacological treatments, such as SSRIs, in managing depression in patients with dementia. Recommended nonpharmacological interventions for managing depression in dementia include decreasing social isolation, behavioral activation (i.e., activities associated with positive mood), exercise programs, and increasing pleasurable activities. However, there are other views on the treatment of depression in dementia. For example, Borisovskaya, Pascualy, and Borson (2014) investigated treatment strategies for cognitive and neuropsychiatric symptoms of AD including agitation, psychosis, anxiety, and depression. Borisovskaya et al. (2014) argue that individualized treatment should provide the best outcome. It is recommended that treatment begins with nonpharmacological interventions, only adding pharmacological treatment when necessary.

Our research also shows that behavior and depression are interlinked, such that managing behavioral impairment can reduce depressive symptoms, and vice versa. Specifically, treatments that attempt to ameliorate cognitive dysfunction in dementia patients tend to also decrease depression, such as reminiscence therapy and CBT. Similarly, managing depression using stress management techniques was also found to ameliorate cognitive decline. Future work should also investigate how managing stress in caregiver of patients of dementia can impact depression in dementia patients (for discussion on this point, see Kor, Liu, & Chien, 2019, 2020).

References

Ahern, E., Kinsella, S., & Semkovska, M. (2018). Clinical efficacy and economic evaluation of online cognitive behavioral therapy for major depressive disorder: A systematic review and meta-analysis. *Expert Review of Pharmacoeconomics & Outcomes Research*, *18*(1), 25–41. Available from https://doi.org/10.1080/14737167.2018.1407245.

Ambrosi, C., Zaiontz, C., Peragine, G., Sarchi, S., & Bona, F. (2019). Randomized controlled study on the effectiveness of animal-assisted therapy on depression, anxiety, and illness perception in institutionalized elderly. *Psychogeriatrics*, *19*(1), 55–64. Available from https://doi.org/10.1111/psyg.12367.

Baek, S. M., Lee, Y., & Sohng, K. Y. (2020). The psychological and behavioural effects of an animal-assisted therapy programme in Korean older adults with dementia. *Psychogeriatrics*, *20*(5), 645–653. Available from https://doi.org/10.1111/psyg.12554.

Barca, M. L., Selbaek, G., Laks, J., & Engedal, K. (2008). The pattern of depressive symptoms and factor analysis of the Cornell Scale among patients in Norwegian nursing homes. *International Journal of Geriatric Psychiatry*, *23*(10), 1058–1065. Available from https://doi.org/10.1002/gps.2033.

Bennett, S., & Thomas, A. J. (2014). Depression and dementia: Cause, consequence or coincidence? *Maturitas*, *79*(2), 184–190. Available from https://doi.org/10.1016/j.maturitas.2014.05.009.

Birks, J. (2006). Cholinesterase inhibitors for Alzheimer's disease. *Cochrane Database of Systemic Reviews* (1), CD005593. Available from https://doi.org/10.1002/14651858.CD005593.

Borisovskaya, A., Pascualy, M., & Borson, S. (2014). Cognitive and neuropsychiatric impairments in Alzheimer's disease: Current treatment strategies. *Current Psychiatry Reports*, *16*(9), 470. Available from https://doi.org/10.1007/s11920-014-0470-z.

Caballero, J., Hitchcock, M., Beversdorf, D., Scharre, D., & Nahata, M. (2006). Long-term effects of antidepressants on cognition in patients with Alzheimer's disease. *Journal of Clinical Pharmacy and Therapeutics*, *31*(6), 593–598. Available from https://doi.org/10.1111/j.1365-2710.2006.00778.x.

Chan, J. Y. C., Chan, T. K., Kwok, T. C. Y., Wong, S. Y. S., Lee, A. T. C., & Tsoi, K. K. F. (2020). Cognitive training interventions and depression in mild cognitive impairment and dementia: A systematic review and meta-analysis of randomized controlled trials. *Age Ageing*, *49*(5), 738–747. Available from https://doi.org/10.1093/ageing/afaa063.

Cheung, D. S. K., Kor, P. P. K., Jones, C., Davies, N., Moyle, W., Chien, W. T., ... Lai, C. (2020). The use of modified mindfulness-based stress reduction and mindfulness-based cognitive therapy programme for family caregivers of people living with dementia: A feasibility study. *Asian Nursing Research (Korean Society of Nursing Science)*, *14*(4)), 221–230. Available from https://doi.org/10.1016/j.anr.2020.08.009.

Cummings, J. L., Miller, B., Hill, M. A., & Neshkes, R. (1987). Neuropsychiatric aspects of multi-infarct dementia and dementia of the Alzheimer type. *Archives of Neurology*, *44*(4), 389–393.

Dafsari, F. S., & Jessen, F. (2020). Depression-an underrecognized target for prevention of dementia in Alzheimer's disease. *Translational Psychiatry*, *10*(1), 160. Available from https://doi.org/10.1038/s41398-020-0839-1.

Evers, M. M., Samuels, S. C., Lantz, M., Khan, K., Brickman, A. M., & Marin, D. B. (2002). The prevalence, diagnosis and treatment of depression in dementia patients in chronic care facilities in the last six months of life. *International Journal of Geriatric Psychiatry*, *17*(5), 464–472. Available from https://doi.org/10.1002/gps.634.

Forstmeier, S., Maercker, A., Savaskan, E., & Roth, T. (2015). Cognitive behavioural treatment for mild Alzheimer's patients and their caregivers (CBTAC): Study protocol for a randomized controlled trial. *Trials*, *16*, 526. Available from https://doi.org/10.1186/s13063-015-1043-0.

Gellis, Z. D., McClive-Reed, K. P., & Brown, E. (2009). Treatments for depression in older persons with dementia. *Annals of Longterm Care*, *17*(2), 29–36.

Giovagnoli, A. R., Manfredi, V., Schifano, L., Paterlini, C., Parente, A., & Tagliavini, F. (2018). Combining drug and music therapy in patients with moderate Alzheimer's disease: A randomized study. *Neurological Sciences*, *39*(6), 1021–1028. Available from https://doi.org/10.1007/s10072-018-3316-3.

Herrmann, N., & Lanctot, K. L. (2007). Pharmacologic management of neuropsychiatric symptoms of Alzheimer disease. *The Canadian Journal of Psychiatry*, *52*(10), 630–646. Available from https://doi.org/10.1177/070674370705201004.

Hopkinson, M. D., Reavell, J., Lane, D. A., & Mallikarjun, P. (2018). Cognitive behavioral therapy for depression, anxiety, and stress in caregivers of dementia patients: A systematic review and meta-analysis. *Gerontologist*, *59*(4), e343–e362. Available from https://doi.org/10.1093/geront/gnx217.

Huang, Y. Y., Dou, K. X., Zhong, X. L., Shen, X. N., Chen, S. D., Li, H. Q., ... Yu, J. T. (2020). Pharmacological treatment of neuropsychiatric symptoms of dementia: A network meta-analysis protocol. *Annals of Translational Medicine*, *8*(12), 746. Available from https://doi.org/10.21037/atm-20-611.

Jung, Y. H., Lee, S., Kim, W. J., Lee, J. H., Kim, M. J., & Han, H. J. (2020). Effect of integrated cognitive intervention therapy in patients with mild to moderate Alzheimer's disease. *Dementia and Neurocognitive Disorders*, *19*(3), 86–95. Available from https://doi.org/10.12779/dnd.2020.19.3.86.

Kor, P. P. K., Liu, J. Y., & Chien, W. T. (2019). Effects on stress reduction of a modified mindfulness-based cognitive therapy for family caregivers of those with dementia: Study protocol for a randomized controlled trial. *Trials*, *20*(1), 303. Available from https://doi.org/10.1186/s13063-019-3432-2.

Kor, P. P. K., Liu, J. Y. W., & Chien, W. T. (2020). Effects of a modified mindfulness-based cognitive therapy for family caregivers of people with dementia: A randomized clinical trial. *Gerontologist*, gnaa125. Available from https://doi.org/10.1093/geront/gnaa125, September 4.

Lam, H. L., Li, W. T. V., Laher, I., & Wong, R. Y. (2020). Effects of music therapy on patients with dementia – A systematic review. *Geriatrics (Basel)*, *5*(4), 62. Available from https://doi.org/10.3390/geriatrics5040062.

Lawton, M. P., Van Haitsma, K., & Klapper, J. (1996). Observed affect in nursing home residents with Alzheimer's disease. *The Journal of Gerontology B Series, Psychological Sciences and Social Sciences*, *51*(1), P3–P14. Available from https://doi.org/10.1093/geronb/51b.1.p3.

Lenze, E. J. (2011). Treating depression in older adults with dementia. *Journal of the American Geriatrics Society*, *59*(4), 754–755. Available from https://doi.org/10.1111/j.1532-5415.2011.03357.x.

Lichtenberg, P. A., Kemp-Havican, J., Macneill, S. E., & Schafer Johnson, A. (2005). Pilot study of behavioral treatment in dementia care units. *Gerontologist*, *45*(3), 406–410.

Linnemann, C., & Lang, U. E. (2020). Pathways connecting late-life depression and dementia. *Frontiers in Pharmacology*, *11*, 279. Available from https://doi.org/10.3389/fphar.2020.00279.

Lyketsos, C. G., DelCampo, L., Steinberg, M., Miles, Q., Steele, C. D., Munro, C., ... Rabins, P. V. (2003). Treating depression in Alzheimer disease: Efficacy and safety of sertraline therapy, and the benefits of depression reduction: The DIADS. *Archives of General Psychiatry*, *60*(7), 737–746. Available from https://doi.org/10.1001/archpsyc.60.7.737.

Marriott, A., Donaldson, C., Tarrier, N., & Burns, A. (2000). Effectiveness of cognitive-behavioural family intervention in reducing the burden of care in carers of patients with Alzheimer's disease. *British Journal of Psychiatry*, *176*, 557–562.

Menon, A. S., Gruber-Baldini, A. L., Hebel, J. R., Kaup, B., Loreck, D., Itkin Zimmerman, S., ... Magaziner, J. (2001). Relationship between aggressive behaviors and depression among nursing home residents with dementia. *International Journal of Geriatric Psychiatry*, *16*(2), 139–146.

Merriam, A. E., Aronson, M. K., Gaston, P., Wey, S. L., & Katz, I. (1988). The psychiatric symptoms of Alzheimer's disease. *Journal of the American Geriatrics Society*, *36*(1), 7–12.

Modrego, P. J., & Ferrandez, J. (2004). Depression in patients with mild cognitive impairment increases the risk of developing dementia of Alzheimer type: A prospective cohort study. *Archives of Neurology*, *61*(8), 1290–1293. Available from https://doi.org/10.1001/archneur.61.8.1290.

Moon, S., & Park, K. (2020). The effect of digital reminiscence therapy on people with dementia: A pilot randomized controlled trial. *BMC Geriatrics*, *20*(1), 166. Available from https://doi.org/10.1186/s12877-020-01563-2.

Moreno-Morales, C., Calero, R., Moreno-Morales, P., & Pintado, C. (2020). Music therapy in the treatment of dementia: A systematic review and meta-analysis. *Frontiers in Medicine (Lausanne)*, 7, 160. Available from https://doi.org/10.3389/fmed.2020.00160.

Mota Pereira, J., & Fonte, D. (2018). Pets enhance antidepressant pharmacotherapy effects in patients with treatment resistant major depressive disorder. *Journal of Psychiatric Research*, *104*, 108–113. Available from https://doi.org/10.1016/j.jpsychires.2018.07.004.

Mowla, A., Mosavinasab, M., Haghshenas, H., & Borhani Haghighi, A. (2007). Does serotonin augmentation have any effect on cognition and activities of daily living in Alzheimer's dementia? A double-blind, placebo-controlled clinical trial. *Journal of Clinical Psychopharmacology*, *27*(5), 484–487. Available from https://doi.org/10.1097/jcp.0b013e31814b98c1.

Nagata, T., Shinagawa, S., Nakajima, S., Noda, Y., & Mimura, M. (2020). Pharmacological management of behavioral disturbances in patients with Alzheimer's disease. *Expert Opinion on Pharmacotherapy*, *21*(9), 1093–1102. Available from https://doi.org/10.1080/14656566.2020.1745186.

Naismith, S. L., & Mowszowski, L. (2016). Moving beyond mood: Is it time to recommend cognitive training for depression in older adults? In B. T. Baune, & P. J. Tully (Eds.), *Cardiovascular diseases and depression* (pp. 365–394). Cham: Springer.

Nelson, J. C., & Devanand, D. P. (2011). A systematic review and meta-analysis of placebo-controlled antidepressant studies in people with depression and dementia. *Journal of the American Geriatrics Society*, *59*(4), 577–585. Available from https://doi.org/10.1111/j.1532-5415.2011.03355.x.

Onega, L. L., Pierce, T. W., & Epperly, L. (2018). Bright light therapy to treat depression in individuals with mild/moderate or severe dementia. *Issues in Mental Health Nursing*, *39*(5), 1–4. Available from https://doi.org/10.1080/01612840.2018.1437648.

Orgeta, V., Qazi, A., Spector, A., & Orrell, M. (2015). Psychological treatments for depression and anxiety in dementia and mild cognitive impairment: Systematic review and meta-analysis. *British Journal of Psychiatry*, *207*(4), 293–298. Available from https://doi.org/10.1192/bjp.bp.114.148130.

Regan, C., Katona, C., Walker, Z., & Livingston, G. (2005). Relationship of exercise and other risk factors to depression of Alzheimer's disease: The LASER-AD study. *International Journal of Geriatric Psychiatry*, *20*(3), 261–268. Available from https://doi.org/10.1002/gps.1278.

Reifler, B. V., Teri, L., Raskind, M., Veith, R., Barnes, R., White, E., & McLean, P. (1989). Double-blind trial of imipramine in Alzheimer's disease patients with and without depression. *The Americal Journal of Psychiatry*, *146*(1), 45–49. Available from https://doi.org/10.1176/ajp.146.1.45.

Rodrigues, J., Capuano, A. W., Barnes, L. L., Bennett, D. A., & Shah, R. C. (2018). Effect of antidepressant medication use and social engagement on the level of depressive symptoms in community-dwelling, older African Americans and Whites with dementia. *Journal of Aging Health*, *31*(7), 1278–1296. Available from https://doi.org/10.1177/0898264318772983, 898264318772983.

Rubin-Falcone, H., Weber, J., Kishon, R., Ochsner, K., Delaparte, L., Dore, B., ... Miller, J. M. (2018). Longitudinal effects of cognitive behavioral therapy for depression on the neural correlates of emotion regulation. *Psychiatry Research: Neuroimaging*, *271*, 82–90. Available from https://doi.org/10.1016/j.pscychresns.2017.11.002.

Sink, K. M., Holden, K. F., & Yaffe, K. (2005). Pharmacological treatment of neuropsychiatric symptoms of dementia: A review of the evidence. *The Journal of the American Medical Assocoiation, 293*(5), 596–608. Available from https://doi.org/10.1001/jama.293.5.596.

Smith-Ray, R. L., Irmiter, C., & Boulter, K. (2016). Cognitive training among cognitively impaired older adults: A feasibility study assessing the potential improvement in balance. *Frontiers in Public Health, 4*, 219. Available from https://doi.org/10.3389/fpubh.2016.00219.

Starkstein, S. E., Chemerinski, E., Sabe, L., Kuzis, G., Petracca, G., Teson, A., & Leiguarda, R. (1997). Prospective longitudinal study of depression and anosognosia in Alzheimer's disease. *British Journal of Psychiatry, 171*, 47–52.

Starkstein, S. E., & Mizrahi, R. (2006). Depression in Alzheimer's disease. *Expert Review of Neurotherapeutics, 6*(6), 887–895. Available from https://doi.org/10.1586/14737175.6.6.887.

Steck, N., Cooper, C., & Orgeta, V. (2018). Investigation of possible risk factors for depression in Alzheimer's disease: A systematic review of the evidence. *Journal of Affective Disorders, 236*, 149–156. Available from https://doi.org/10.1016/j.jad.2018.04.034.

Steffens, D. C., Skoog, I., Norton, M. C., Hart, A. D., Tschanz, J. T., Plassman, B. L., ... Breitner, J. C. (2000). Prevalence of depression and its treatment in an elderly population: The Cache County study. *Archives of General Psychiatry, 57*(6), 601–607.

Stewart, D. B., Berg-Weger, M., Tebb, S., Sakamoto, M., Roselle, K., Downing, L., ... Hayden, D. (2017). Making a difference: A study of cognitive stimulation therapy for persons with dementia. *Journal of Gerontological Social Work, 60*(4), 300–312. Available from https://doi.org/10.1080/01634372.2017.1318196.

Sukontapol, C., Kemsen, S., Chansirikarn, S., Nakawiro, D., Kuha, O., & Taemeeyapradit, U. (2018). The effectiveness of a cognitive training program in people with mild cognitive impairment: A study in urban community. *Asian Journal of Psychiatry, 35*, 18–23. Available from https://doi.org/10.1016/j.ajp.2018.04.028.

Takemoto, M., Ohta, Y., Hishikawa, N., Yamashita, T., Nomura, E., Tsunoda, K., ... Abe, K. (2020). The efficacy of sertraline, escitalopram, and nicergoline in the treatment of depression and apathy in Alzheimer's disease: The Okayama depression and apathy project (ODAP). *Journal of Alzheimer's Disease, 76*(2), 769–772. Available from https://doi.org/10.3233/JAD-200247.

Teng, E., Ringman, J. M., Ross, L. K., Mulnard, R. A., Dick, M. B., Bartzokis, G., ... Alzheimer's Disease Research Centers of California-Depression in Alzheimer's Disease. (2008). Diagnosing depression in Alzheimer disease with the National Institute of Mental Health provisional criteria. *American Journal of Geriatric Psychiatry, 16*(6), 469–477. Available from https://doi.org/10.1097/JGP.0b013e318165dbae.

Teri, L., & Gallagher-Thompson, D. (1991). Cognitive-behavioral interventions for treatment of depression in Alzheimer's patients. *Gerontologist, 31*(3), 413–416.

Teri, L., Gibbons, L. E., McCurry, S. M., Logsdon, R. G., Buchner, D. M., Barlow, W. E., ... Larson, E. B. (2003). Exercise plus behavioral management in patients with Alzheimer disease: a randomized controlled trial. *The Journal of the American Medical Assocoiation, 290*(15), 2015–2022. Available from https://doi.org/10.1001/jama.290.15.2015.

Teri, L., Logsdon, R. G., Uomoto, J., & McCurry, S. M. (1997). Behavioral treatment of depression in dementia patients: a controlled clinical trial. *The Journal of Gerontology B Series, Psychological Sciences and Social Sciences, 52*(4), P159–P166.

Teri, L., Reifler, B. V., Veith, R. C., Barnes, R., White, E., McLean, P., & Raskind, M. (1991). Imipramine in the treatment of depressed Alzheimer's patients: impact on cognition. *Journal of Gerontology, 46*(6), P372–P377.

Vara, M. D., Herrero, R., Etchemendy, E., Espinoza, M., Banos, R. M., Garcia-Palacios, A., ... Botella, C. (2018). Efficacy and cost-effectiveness of a blended cognitive behavioral therapy for depression in Spanish primary health care: study protocol for a randomised non-inferiority trial. *BMC Psychiatry*, *18*(1), 74. Available from https://doi.org/10.1186/s12888-018-1638-6.

Walker, D. A. (2004). Cognitive behavioural therapy for depression in aperson with Alzheimer's disease. *Behavioural and Cognitive Psychotherapy*, *32*, 495–500.

Williams, C. L., & Tappen, R. M. (2007). Effect of exercise on mood in nursing home residents with Alzheimer's disease. *American Journal of Alzheimer's Disease & Other Dementias*, *22*(5), 389–397. Available from https://doi.org/10.1177/1533317507305588.

Williams, C. L., & Tappen, R. M. (2008). Exercise training for depressed older adults with Alzheimer's disease. *Aging and Mental Health*, *12*(1), 72–80. Available from https://doi.org/10.1080/13607860701529932.

Woods, B., O'Philbin, L., Farrell, E. M., Spector, A. E., & Orrell, M. (2018). Reminiscence therapy for dementia. *Cochrane Database of Systemic Reviews* (3), CD001120. Available from https://doi.org/10.1002/14651858.CD001120.pub3.

Zhang, Z., Zhang, L., Zhang, G., Jin, J., & Zheng, Z. (2018). The effect of CBT and its modifications for relapse prevention in major depressive disorder: A systematic review and meta-analysis. *BMC Psychiatry*, *18*(1), 50. Available from https://doi.org/10.1186/s12888-018-1610-5.

CHAPTER 10

Developing a Vulnerability to Negative Affect in Dementia Scale (VNADS) for music interventions

Sandra Garrido[1], Wafa Jaroudi[2] and Ahmed A. Moustafa[1,3,4]
[1]School of Psychology, Western Sydney University, Sydney, NSW, Australia
[2]School of Social Sciences and Psychology, Western Sydney University, Sydney, NSW, Australia
[3]MARCS Institute for Brain and Behaviour, Western Sydney University, Sydney, NSW, Australia
[4]Department of Human Anatomy and Physiology, The Faculty of Health Sciences, University of Johannesburg, Johannesburg, South Africa

Introduction

As many as 20% of people with Alzheimer's disease (AD) and 30% of people with dementia with Lewy bodies may also have major depressive disorder (MDD) (Whitfield et al., 2015). Similarly, 33% of people with frontotemporal dementia have depressive symptoms (Chakrabarty, Sepehry, Jacova, & Hsiung, 2015), and 63% of people with mild cognitive impairment (MCI) may also have MDD (Panza et al., 2010). Along with the high rates of comorbidity between depression and dementia, there is also strong evidence of a bi-directional relationship. Depressive episodes in later life can increase the risk of developing dementia by up to three times (Brommelhoff et al., 2009), and people with MCI are twice as likely to develop AD if they also have MDD (Ownby, Crocco, Acevedo, John, & Loewensteing, 2006). In addition, the presence of depression seems to increase the rates of cognitive and functional decline as well as hasten institutionalization in people with dementia (Potter & Steffens, 2007).

Given the frequent cooccurrence of depression and dementia, some believe that the neuropsychological changes that occur with recurrent depressive episodes increase the brain's vulnerability to the degenerative processes involved in dementia (Aznar & Knudsen, 2011). The close connection between the two diseases has led some to argue that a 25% reduction in depression prevalence could result in 827,000 fewer cases of AD worldwide (Bames & Yaffe, 2011). Depression also has a substantial impact on quality of

Alzheimer's Disease
DOI: https://doi.org/10.1016/B978-0-12-821334-6.00007-7

life (Banerjee et al., 2009), compounding the stress upon both caregivers and health services (Curran & Loi, 2012). Agitation is also common in people with dementia, encompassing a range of symptoms such as repetitive acts, restlessness, wandering, and aggression (Cerejeira, Lagarto, & Mukaetova-Ladinska, 2012). It occurs in up to 80% of people with dementia (Sourial, McCusker, Cole, & Abrahamowicz, 2001) and imposes increased stress on both patients and caregivers, increasing disability and diminishing the quality of life (Kales, Gitlin, & Lyketsos, 2014).

The presence of multiple comorbidities and high levels of medication being prescribed for other conditions makes pharmacological approaches to addressing the behavioral and psychological symptoms of dementia problematic in this population (Gottfries, 2001). The antidepressants, anxiolytics, and antipsychotics often used are associated with side effects including nausea and insomnia (Caltagirone et al., 2005). Evidence suggests that they are not always effective in diminishing symptoms and can be associated with severe adverse clinical events such as increased confusion (Farina, Morrell, & Banerjee, 2017), or even death (Sacchetti, Turrina, & Valsecchi, 2010). Thus nonpharmacological treatments are often recommended as a first-line treatment for psychological and behavioral symptoms (Arzemia, Elseviers, Petrovic, Van Bortel, & Stichele, 2011). Despite this, pharmacological treatments are still widely used, partly due to a lack of standardized and evidence-based guidelines around nonpharmacological solutions (Hansen, Gartlehner, Lohr, & Kaufer, 2007).

Research demonstrates that music can have a positive effect on the quality of life and mood in people with dementia. Numerous studies have demonstrated that active forms of music therapy can have a positive influence on depressive symptoms, agitation, and cognitive functioning in people with dementia (O'Connor, Ames, Gardner, & King, 2009; Van der Steen et al., 2018). Raglio and colleagues (Raglio et al., 2015), for example, demonstrated that music therapy supports the management of agitation in individuals with dementia.

Despite these obvious benefits, musical interventions for people with dementia that require the presence of a professional therapist or live musician are limited by significant costs and logistical factors (Nair, Browne, Marley, & Heim, 2013). Thus while many aged care facilities attempt to implement music programs of some sort, they are often unable to utilize programs by professional music therapists as often as desirable (Garrido, Dunne, Perz, Chang, & Stevens, 2018a; Garrido, Stevens, Chang, Dunne, & Perz, 2018b). There is thus increasing interest in the use of prerecorded music for therapeutic purposes outside of formal music therapy settings.

Such interventions have the advantage of being relatively easy to access and affordable. Prerecorded music is widely available to most people whether in residential or home care situations and can be accessed at the time it is most needed, as frequently as needed. There is an increasing recognition that self-selected music and music that is biographically salient are more effective as mood moderators than music that an individual has no particular affinity with and may even actively dislike (Nair et al., 2013). Thus interest has begun to shift toward the use of personalized playlists in aged care contexts. "Personalized playlists" refer to the creation of playlists of prerecorded music based on the tastes of the individual listener.

However, research suggests that using music with vulnerable populations such as people with dementia requires more care than merely selecting an individual's "favorite music." Some studies, for example, while reporting that a majority of participants benefited from the use of personalized playlists based on favorite music, acknowledged that positive results were not universal, with some participants responding neutrally or even experiencing a worsening of mood (Martin, Schroeder, Smith, & Jones, 2016; Ziv, Granot, Hai, Dassa, & Haimov, 2007). Research has demonstrated that both the features of the music such as tempo and mode (Garrido, Stevens, Chang, Dunne, & Perz, 2019) and individual variables such as having a history of recent or lifelong depression and the severity of their dementia (Garrido et al., 2018a, 2018b) can influence the affective outcome of personalized playlists. Other studies have previously shown that personality factors such as rumination (Garrido, Bangert, & Schubert, 2016) could be predictive of undesirable mood outcomes when listening to music in healthy participants. As outlined previously, the literature also suggested that agitation could be increased in some circumstances (Garland, Beer, Eppingstall, & O'Connor, 2007) and that strong memories and overwhelming emotions in response to music could be triggered in people with dementia (Moore, 2014).

These findings demonstrate the importance of assessing the suitability of potential candidates for music interventions and determines individual cases in which such programs should be implemented with particular care. This is especially important in the case of music programs that do not involve the use of a trained music therapist, such as where nursing staff or volunteers may wish to utilize music programs for residents with dementia. While experiencing negative emotions in response to music is not always undesirable and can even have therapeutic benefits at times, research shows that such episodes can be detrimental for individuals with a history of depression in supported situations (Garrido & Schubert, 2015). Music therapists are trained in responding to adverse responses that can occur in relation to music such as overstimulation, emotional flooding, or the triggering of distressing memories (Moore, 2014). However, untrained caregivers may be less equipped to deal with these negative responses. There is therefore a need to identify at-risk individuals before implementing music interventions in people with dementia so as to ensure that appropriate monitoring and support systems are in place.

To date there are no tools available to help care staff in aged care facilities or music therapists identify individuals with whom particular care may need to be taken. While the factors discussed previously could help identify vulnerable participants, assessing potential participants of music programs on these factors using previously established scales in their entirety would result in considerable burden to the caregiver. The current study aimed to address this gap by developing and evaluating a screening tool for identifying individuals who may be vulnerable to adverse responses to music interventions.

Method

Item development

The first set of items for the Vulnerability to Negative Affect in Dementia Scale (VNADS) was formulated based on the literature and established scales and designed to reflect the factors that had previously been discovered to be associated with negative responses to music. Therefore as a first step to scale development we compiled a set of items drawn partly from preestablished scales such as the Generalized Anxiety Disorder Scale (Kroenke, Lowe, Spitzer, & Williams, 2006), the Patient Health Questionnaire (Spitzer, Kroenke, & Williams, 1999), the Rumination, Reflectiveness Questionnaire (Trapnell & Campbell, 1999), and two items developed specifically for the VNADS. Strongest loading items from these previously developed scales were used after eliminating the items that were conceptually duplicated across scales. The final list included the 24 items that were rated on a scale of 1 (Strongly disagree) to 5 (Strongly Agree).

Pretesting the items

Version 1 of the VNADS was first pretested with 384 cognitively healthy participants in order to explore the structure and validity of the resulting 24-item scale. A sample from the general population was selected rather than people with dementia for the pretest, since this allowed us to explore the general structure of the scale with a large sample before testing it with people with dementia. Participants were aged 18 years of age and above and were recruited to participate in an online survey via promotion on various websites, blogs, discussion pages, and social media platforms. Other scales included for validation purposes are the Leiden Index of Depressive Sensitivity (Van der Does & Williams, 2003), the Geriatric Anxiety Inventory (GAI) (Pachana et al., 2007), the 21-item Depression Anxiety Stress Scale (DASS-21) (Lovibond & Lovibond, 1995), and the Kessler 10 (K10) (Kessler et al., 2003). Correlation analyses to examine relationships with previous scales were conducted along with a principal components analysis and reliability analyses.

Piloting the VNADS (proxy) with people with dementia

Based on the results of the pretest, a refined version of the VNADS was created and an alternative version designed to be completed by family or carers was also created. This proxy version had the wording adapted to reflect the individual's observations of the past or current behavior or affective states of the person with dementia. The proxy version was tested with a sample of 48 participants (18 males and 30 females) with dementia aged 60 and over (mean = 64.9, SD = 2.7) and their carers in an online survey that included an embedded simulation of a music-listening session. Participant dyads were recruited via advertisements in relevant websites and social media pages and distribution to the mailing lists of various dementia support groups in NSW, Australia. Participants with dementia had been diagnosed with Alzheimer's dementia (48.1%), had a nonspecific diagnosis of dementia (13.5%), vascular dementia (7.7%), frontotemporal dementia (3.8%), MCI (3.8%), or other types of dementia (23.1%). The level of cognitive impairment was assessed by the Dementia Severity Rating Scale (DSRS) (Clark & Ewbank, 1996) for proxy completion. After the completion of the DSRS and the VNADS (proxy version), the survey then directed participants to select three songs from a list of five songs. The songs were from a number of genres thought likely to match the taste of the age group of the participants namely, classical, country, folk, popular, and jazz. The selected songs were designed to evoke sadness in the listener, being in a minor key and relatively slow tempos as is typical of sad music in Western musical traditions. After answering a question to confirm that they had been able to listen to the song, the participants (or their carer) were given the instruction: "How did listening to this song make you feel (or the person with dementia if you are filling this out on their behalf)? Move the slider to show the effect of the music." The participant then rated the effect of the music on a slider that could be moved from negative (1) to positive (5) with neutral as a midterm. They also rated the song's level of familiarity, since this can also play a role in response to music (Sung, Chang, & Lee, 2010). In order to counteract any negative effect of listening to such music, participants were then able to select a song from a list of happier songs before concluding the survey, and similarly rated their response to it. Scale structure was again explored through factor analyses and its usefulness in identifying individuals likely to have a negative affective response to music.

Results

Pretest of version 1

The VNADS items were subjected to a principal components analysis using oblimin rotation with Kaiser normalization. Four components had eigenvalues >1. However, the screen plot suggested a model dominated by one primary component and three less-influential components. The last component was not easily interpreted since it was made up of items that also loaded strongly onto the first component. Thus we retained three components in the final model, which cumulatively explained 56.6% of the variance. Therefore it was determined that the scale represented three components, which were labeled (1) depression, (2) rumination, and (3) emotionality. A reliability analysis returned a Cronbach's alpha of 0.93 for the scale as a whole, which would improve marginally to 0.94 with the removal of Item 22. Therefore this item was removed from the scale.

Scale totals were calculated for the remaining 23 items of the VNADS and for the other scales according to authors' instructions. Pearson correlations were calculated in order to determine whether the VNADS had strong association with other scales, measuring factors likely to be predictive of negative response to music programs. The VNADS was positively correlated with the LIEDS ($r=69$, $P<.001$), the DASS Stress subscale ($r=69$, $P<.001$), the DASS Anxiety subscale ($r=62$, $P<.001$), the DASS Depression subscale ($r=71$, $P<.001$), the K10 ($r=78$, $P<.001$), and the GAI ($r=74$, $P<.001$).

Pilot of the VNADS (proxy)

A confirmatory factor analysis of the 23-item VNADS (proxy) items was performed using the unweighted least squares method of extraction with varimax rotation with a fixed extraction of three factors. Items with communalities below 5 and with cross-loadings of a difference less than 2

Table 10.1 Factor loadings for revised VNADS (proxy) items.

Item no.		Depression	Rumination	Emotionality
1	In the last 2 weeks he/she has felt down, gloomy or hopeless	0.745		0.499
2	Often throughout his/her life he/she has felt down, depressed or hopeless	0.838		
3	In the past 2 weeks he/she has felt like he/she is a failure or has let others down	0.764		
4	Often throughout his/her life he/she has felt bad about him/herself, like he/she is a failure or have let others down	0.802		
5	In the past 2 weeks he/she has been unable to stop worrying			0.577
6	He/she often plays back in his/her mind how he/she acted in a past situation		0.853	
7	He/she often re-evaluates something he/she had done in the past		0.785	
8	He/she often reflects on episodes of his/her life that he/she should no longer be concerned with		0.895	
9	He/she finds it hard to avoid thinking about traumatic memories of previous events in his/her life		0.678	
10	He/she becomes tearful or appears to feel like crying more often than he/she used to			0.689

were removed leaving 10 items in the scale (Table 10.1). The factor analysis was performed again returning a three-factor solution, which were labeled (1) depression, (2) rumination, and (3) emotionality. These three factors explained 69.66% of the variability in the data. Reliability analysis of those 10 items revealed a Cronbach's alpha of 89.

Participants were then categorized as either No to Low Risk or Moderate to High Risk based on scores on this 10-item scale. Since a score of 3 on each item was a rating of "Neither Agree nor Disagree," participants were categorized as No to Low Risk if they score 33 or under, that is, agreed to no more than one item on the scale. They were categorized as Moderate to High Risk if they score 34 or above, that is, agreed to 2 or more items on the scale.

A Mann–Whitney test was then calculated with response to sad music as the dependent variable and VNADS Risk Group as the independent variable. Results indicated that the No to Low Risk group rated higher (more positive) mood responses to the music ($n = 31$, median $= 4$) than the Moderate to High Risk group ($n = 10$, median $= 2$). These results were statistically significant $U = 94$, $z = -1.98$, $P = .044$. A similar analysis was performed with response to happy music as the dependent variable, but there were no significant differences between groups. These results suggest that the 10-item form of the VNADS (proxy) was useful in identifying music listeners with dementia at risk for negative affective responses to music, in particular sad music.

Discussion

The two studies reported in this chapter aimed to develop and conduct preliminary testing of a short questionnaire that could identify people with dementia who were likely to experience negative affective responses to music. The first study tested a 24-item questionnaire with cognitively healthy participants, finding high internal reliability and moderate to strong correlations with longer, well-validated scales measuring similar aspects. In the second study, a shortened (10-item) proxy version of the VNADS was able to identify individuals likely to be at risk for negative responses to music. Items in this short version loaded onto three related dimensions: (1) depression, (2) rumination, and (3) emotionality.

These findings align with previous studies of both people with dementia (Garrido et al., 2018a, 2018b; Martin et al., 2016) and younger adults (Chen, Zhou, & Bryant, 2007; Garrido & Schubert, 2015; Wilhelm, Gilllis, Schubert,

& Whittle, 2013) that have found that people with depression or people who are high ruminators are prone to experiencing negative affective outcomes when listening to sad music. While the longer version of the VNADS also contained items relating to the intensity of emotional responses, these items had weaker factor loadings, suggesting that being prone to intense emotional responses to music or other events or stimuli is less predictive of adverse reactions to music than overall mental health and lifelong thinking patterns such as rumination. Rumination, which is itself closely associated with depression, is characterized by compulsive thinking about negative things (Nolen-Hoeksema & Morrow, 1993; Trapnell & Campbell, 1999). People who are ruminators or who experience episodes of major depression tend to have an attentional bias toward negative stimuli (Gotlib, Krasnoperova, Neubauer Yue, & Joormann, 2004; Platt, Murphy, & Lau, 2015) and difficulty recovering from negative affective states (Foland-Ross et al., 2013). Thus it is logical that such individuals may be more prone to experiencing adverse reactions when listening to sad music. In contrast, intense emotional responses may be related to psychological mechanisms such as catharsis which, while negative initially, ultimately have positive psychological benefits (Garrido & Schubert, 2013), even reducing ruminative thinking (Zhan et al., 2017). This suggests that intense emotional responses to music do not indicate that engagement with music is detrimental to the individual. Even for those individuals prone to negative responses to "sad" music, benefits can still be obtained by listening to music. However identifying such individuals can enable caregivers to select music more cautiously and be prepared to manage any negative responses.

The studies reported herein are limited by the fact that they were not tested in controlled conditions. Study 2 used a simulated experimental methodology and involved a relatively small sample. Since it was not conducted in a laboratory setting, it is possible that other factors outside of the researcher's control influenced the results. Future studies should look at testing the scale in face-to-face settings, using multiple measurement methodologies to assess the impact of music listening and with a larger sample of people with dementia and their carers. In addition, the experimental paradigm used did not allow for music listening sessions to be fully individualized. While the music choices were designed to allow some personal selection reflecting genre preferences, previous research has shown that music that is personally relevant can have more positive responses (Gerdner, 2012). Furthermore, the study was also limited in the use of proxy measures. Studies have revealed that there are often discrepancies between self-report and proxy measures (Howland, Allan, & Sajatovic,

2017). Finding gold standard methods for measuring psychological constructs in people with dementia is an ongoing challenge in research of this nature. Future studies could consider comparing proxy with self-report versions of the VNADS in people with low to moderate levels of impairment or use behavioral measures for comparative purposes.

Nevertheless, the current study reports on a valuable tool that can help identify individuals needing additional monitoring or support when engaging in music or other arts-based programs. The final 10-item version of the VNADS is unlikely to add to questionnaire fatigue or user burden and thus provides a useful assessment tool for use by care staff or researchers in future studies.

Funding

This research was funded by an NHMRC-ARC Dementia Research Development Fellowship to the first author.

Conflict of interests

The authors declare that there is no conflict of interest regarding the publication of this chapter.

Acknowledgments

None.

References

Arzemia, M., Elseviers, M., Petrovic, M., Van Bortel, L., & Stichele, R. V. (2011). Geriatric drug utilisation of psychotropics in Belgian nursing homes. *Human Psychopharmacology*, *26*, 12–20. Available from https://doi.org/10.1002/hup.1160.

Aznar, S., & Knudsen, G. M. (2011). Depression and Alzheimer's disease: Is stress the initiating factor in a common neuropathological cascade? *Journal of Alzheimer's Disease*, *23*, 177–193.

Bames, D. E., & Yaffe, K. (2011). The projected effect of risk factor reduction on Alzheimer's disease prevalence. *Lancet Neurology*, *10*, 819–828.

Banerjee, S., Samsi, K., Petrie, C. D., Alvir, J., Treglia, M., Schwam, E. M., & del Valle, M. (2009). What do we know about quality of life in dementia? A review of the emerging evidence on the predictive and explanatory value of disease specific measures of health related quality of life in people with dementia. *International Journal of Geriatric Psychiatry*, *24*(1), 15–24.

Brommelhoff, J., Gatz, M., Johansson, B., McArdle, J., Fratiglioni, L., & Pedersen, N. (2009). Depression as a risk factor or prodromal feature for dementia? Findings in a population-based sample of Swedish twins. *Psychology and Aging, 24*, 373–384.

Caltagirone, C., Bianchetti, A., Di Luca, M., Mecocci, P., Padovani, A., Pirfo, E., ... Musicco, M. (2005). Guidelines for the treatment of Alzheimer's disease from the Italian Association of Psychogeriatrics. *Drugs & Aging, 22*(Suppl. 1), 1–26.

Cerejeira, J., Lagarto, L., & Mukaetova-Ladinska, E. B. (2012). Behavioral and psychological symptoms of dementia. *Frontiers in Neurology, 3*(73). Available from https://doi.org/10.3389/fneur.2012.00073.

Chakrabarty, T., Sepehry, A. A., Jacova, C., & Hsiung, G.-Y. R. (2015). The prevalence of depressive symptoms in frontotemporal dementia: A meta-analysis. *Dementia and Geratric Cognitive Disorders, 39*, 257–271.

Chen, L., Zhou, S., & Bryant, J. (2007). Temporal changes in mood repair through music consumption: Effects of mood, mood salience, and individual differences. *Media Psychology, 9*, 695–713.

Clark, C. M., & Ewbank, D. C. (1996). Performance of the dementia severity rating scale: A caregiver questionnaire for rating severity in Alzheimer disease. *Alzheimer Disease and Associated Disorders, 10*(1), 31–39.

Curran, E. M., & Loi, S. (2012). Depression and dementia. *Medical Journal of Australia, 1* (Suppl 4), 40–44.

Farina, N., Morrell, L., & Banerjee, S. (2017). What is the therapeutic value of antidepressants in dementia? A narrative review. *International Journal of Geriatric Psychiatry, 32*(1), 32–49.

Foland-Ross, L. C., Hamilton, J. P., Joormann, J., Berman, M. G., Jonides, J., & Gotlib, I. H. (2013). The neural basis of difficulties disengaging from negative irrelevant material in major depression. *Psychological Science, 24*(3), 334–344.

Garland, K., Beer, E., Eppingstall, B., & O'Connor, D. W. (2007). A comparison of two treatments of agitated behaviour in nursing home redidents with dementia: Simulated family presence and preferred music. *American Journal of Geriatric Psychiatry, 15*(6), 514–521.

Garrido, S., & Schubert, E. (2013). Adaptive and maladaptive attraction to negative emotion in music. *Musicae Scientiae, 17*(2), 145–164. Available from https://doi.org/10.1177/1029864913478305.

Garrido, S., & Schubert, E. (2015). Music and people with tendencies to depression. *Music Perception, 32*(4), 313–321. Available from https://doi.org/10.1525/MP.2015.32.4.313.

Garrido, S., Bangert, D., & Schubert, E. (2016). Musical prescriptions for mood improvements: A mixed methods study. *The Arts in Psychotherapy, 51*, 46–53.

Garrido, S., Dunne, L., Perz, J., Chang, E., & Stevens, C. (2018a). The use of music in aged care facilities: A mixed methods study. *Journal of Health Psychology, 15*, 765–776.

Garrido, S., Stevens, C., Chang, E., Dunne, L., & Perz, J. (2018b). Music and dementia: Individual differences in response to personalized playlists. *Journal of Alzheimer's Disease, 64*(3), 933–941. Available from https://doi.org/10.3233/JAD-180084.

Garrido, S., Stevens, C., Chang, E., Dunne, L., & Perz, J. (2019). Music and dementia: Musical features and affective responses to personalized playlists. *American Journal of Alzhiemers and Other Dementias, 34*(4), 247–253. Available from https://doi.org/10.1177/1533317518808011.

Gerdner, L. A. (2012). Individualized music for dementia: Evolution and application of evidence-based protocol. *World Journal of Psychiatry, 2*(2), 26–32.

Gotlib, I. H., Krasnoperova, E., Neubauer Yue, D., & Joormann, J. (2004). Attentional biases for negative interpersonal stimuli in clinical depression. *Journal of Abnormal Psychology, 113*(1), 127–135.

Gottfries, C.-G. (2001). Late life depression. *European Archives of Psychiatry and Clinical Neuroscience, 251*(Suppl. 2), 57–61.
Hansen, R. A., Gartlehner, G., Lohr, K. N., & Kaufer, D. I. (2007). Functional outcomes of drug treatment in Alzheimer's disease. *Drugs & Aging, 24*, 155–167. Available from https://doi.org/10.2165/00002512-200724020-00007.
Howland, M., Allan, K. C., & Sajatovic, M. (2017). Patient-rated versus proxy-rated cognitive and functional measures in older adults. *Patient Related Outcome Measures, 8*, 33–42. Available from https://doi.org/10.2147/PROM.S126919.
Kales, H. C., Gitlin, L. N., & Lyketsos, C. G. (2014). Management of neuropsychiatric symptoms of dementia is clinical settings: Recommendations from a multidisciplinary panel. *Journal of American Geriatric Society, 62*, 762–769.
Kessler, R. C., Barker, P. R., Colpe, L. J., Epstein, J. F., Gfroerer, J. C., Hiripi, E., ... Zaslavsky, A. M. (2003). Screening for serious mental illness in the general population. *Archives of General Psychiatry, 60*(2), 184–189.
Kroenke, K., Lowe, B., Spitzer, R. L., & Williams, J. B. (2006). A brief measure for assessing generalized anxiety disorder: The GAD-7. *Archives of Internal Medicine, 166*(10), 1092–1097.
Lovibond, S. H., & Lovibond, P. F. (1995). The structure of negative emotinal states: Comparison of the Depression Anxiety Stress Scales (DASS) with the beck depression and anxiety inventories. *Behaviour Research and Therapy, 33*(3), 335–343. Available from https://doi.org/10.1016/0005-7967(94)00075-U.
Martin, P. K., Schroeder, R. W., Smith, J. M., & Jones, B. (2016). The Roth project – Music and memory: Surveying the observed benefit of personalized music in individuals with diagnosed or suspected dementia. *Alzheimer's & Dementia, 12*(7 Suppl.), P988. Available from https://doi.org/10.1016/j.jalz.2016.06.2028.
Moore, K. S. (2014). *Five problems music can create*. Retrieved from https://www.psychology-today.com/us/blog/your-musical-self/201408/5-problems-music-can-create; Accessed 28.11.19.
Nair, B. R., Browne, W., Marley, J., & Heim, C. (2013). Music and dementia. *Degenerative Neurological and Neuromuscular Disease, 3*, 47–51.
Nolen-Hoeksema, S., & Morrow, J. (1993). Effects of rumination and distraction on naturally occurring depressed mood. *Cognition & Emotion*, 7(6), 561–570.
O'Connor, D. W., Ames, D., Gardner, B., & King, M. (2009). Psychosocial treatments of behavior symptoms in dementia: A systematic review of reports meeting quality standards. *International Psychogeriatrics, 21*(2), 225–240.
Ownby, R., Crocco, E., Acevedo, A., John, V., & Loewensteing, D. (2006). Depression and risk for Alzheimer disease: Systematic review, *meta*-analysis, and metaregression analysis. *Archives of General Psychiatry, 63*, 530–538.
Pachana, N., Byrne, G., Siddle, H., Koloski, N., Harley, E., & Arnold, E. (2007). Development and validation of the geriatric anxiety inventory. *International Psychogeriatrics, 19*, 103–114.
Panza, F., Frisardi, V., Capurso, C., D'Introno, A., Colacicco, A., Imbimbo, B., ... Solfrizzi, V. (2010). Late-life depression, mild cognitive impairment, and dementia: Possible continuum? *American Journal of Geriatric Psychiatry, 18*, 98–116.
Platt, B., Murphy, S. E., & Lau, J. Y. F. (2015). The association between negative attentional biases and symptoms of depression in a community sample of adolescents. *PeerJ, 3*, e1372.
Potter, G., & Steffens, D. (2007). Contribution of depression to cognitive impairment and dementia in older adults. *The Neurologist, 13*, 105–117.
Raglio, A., Bellandi, D., Baiardi, P., Gianotti, M., Ubezio, M. C., Zanacchi, E., ... Stramba-Badiale, M. (2015). Effect of active music therapy and individualized listening to music on dementia: A multicenter randomized controlled trial. *Journal of the*

American Geriatrics Society, *63*(8), 1534–1539. Available from https://doi.org/10.1111/jgs.13558.
Sacchetti, E., Turrina, C., & Valsecchi, P. (2010). Cerebrovascular accidents in elderly people treated with antipsychotic drugs. *Drug Safety*, *33*, 273–288.
Sourial, R., McCusker, J., Cole, M., & Abrahamowicz, M. (2001). Agitation in demented patients in an acute care hospital: Prevalence, disruptiveness and staff burden. *International Psychogeriatrics*, *13*, 183–197. Available from https://doi.org/10.1017/S1041610201007578.
Spitzer, R. L., Kroenke, K., & Williams, J. B. (1999). Patient health questionnaire study group. Validity and utility of a self-report version of PRIME-MD: The PHQ primary care study. *JAMA: The Journal of the American Medical Association*, *282*, 1737–1744.
Sung, H. C., Chang, A. M., & Lee, W. L. (2010). A preferred music listening intervention to reduce anxiety in older adults with dementia in nursing homes. *Journal of Clinical Nursing*, *19*(7–8), 1056–1064. Available from https://doi.org/10.1111/j.1365-2702.2009.03016.x.
Trapnell, P. D., & Campbell, J. D. (1999). Private self-consciousness and the five-factor model of personality: Distinguishing rumination from reflection. *Journal of Personality and Social Psychology*, *76*(2), 284–304.
Van der Does, A. J. W., & Williams, J. W. G. (2003). *Leiden Index of Depressive Sensitivity - Revised (LEIDS-R)*. Leiden University. Retrieved from https://www.dousa.nl/publications_depression.htm#LEIDS
Van der Steen, J. T., Smaling, H. J. A., van der Wouden, J. C., Bruinsma, M. S., Scholten, R. J. P. M., & Vink, A. C. (2018). Music-based therapeutic interventions for people with dementia. *Cochrane Database of Systematic Reviews* (7), CD00347. Available from https://doi.org/10.1002/14651858.CD003477.pub4.
Whitfield, D. R., Vallortigara, J., Alghamdi, A., Hortobagyi, T., Ballard, C., Thomas, A. J., ... Francis, P. T. (2015). Depression and synaptic zinc regulatino in Alzheimer disease, dementia with Lewy bodies, and Parkinson disease dementia. *The American Journal of Geriatric Psychiatry*, *23*(2), 141–148.
Wilhelm, K., Gilllis, I., Schubert, E., & Whittle, E. L. (2013). On a blue note: Depressed people's reasons for listening to music. *Music and Medicine*, *5*(2), 76–83.
Zhan, J., Tang, F., Mei, H., Jin, F., Jing, X., Chang, L., & Jing, L. (2017). Regulating rumination by anger: Evidence for the mutual promotion and counteraction (MPMC) theory of emotionality. *Frontiers in Psychology*, *8*, 1871. Available from https://doi.org/10.3389/fpsyg.2017.01871.
Ziv, N., Granot, A., Hai, S., Dassa, A., & Haimov, I. (2007). The effect of background stimulative music on behaviour in Alzheimer's patients. *Journal of Music Therapy*, *44*(4), 329–343.

CHAPTER 11

Using music to improve mental health in people with dementia

Ahmed A. Moustafa[1,2,3], Eid Abo Hamza[4,5], Wafa Jaroudi[6] and Sandra Garrido[1]
[1]School of Psychology, Western Sydney University, Sydney, NSW, Australia
[2]MARCS Institute for Brain and Behaviour, Western Sydney University, Sydney, NSW, Australia
[3]Department of Human Anatomy and Physiology, The Faculty of Health Sciences, University of Johannesburg, Johannesburg, South Africa
[4]Faculty of Education, Department of Mental Health, Tanta University, Tanta, Egypt
[5]College of Graduate Studies, Arabian Gulf University, Bahrain
[6]School of Social Sciences and Psychology, Western Sydney University, Sydney, NSW, Australia

Introduction

Music therapy is an evidence-based approach to improve the quality of life through listening to music (Garrido et al., 2017), as well as more active engagement with music such as singing (Lam, Li, Laher, & Wong, 2020; Perez-Eizaguirre & Vergara-Moragues, 2020), playing musical instruments (Lam et al., 2020; Leggieri et al., 2019; Perez-Eizaguirre & Vergara-Moragues, 2020), and songwriting (Clark, Stretton-Smith, Baker, Lee, & Tamplin, 2020). The practice of music therapy involves a trained and registered music therapist who uses established therapeutic practices to tailor a program of musical "treatment" to the needs of the individual with dementia. Music therapy can improve both physical and language abilities through activities such as singing (Davidson & Garrido, 2015) as well as improving social communication between individuals, allowing them to connect to the songs that have personal meanings (Ridder, 2011). Moreover, music therapy has been used in individuals with dementia to improve psychological well-being by alleviating symptoms of depression (Chu et al., 2014; Lam et al., 2020) and agitation (Garrido et al., 2017). Consequently, music therapy also improves cognitive dysfunction and memory decline of individuals with dementia (Chu et al., 2014).

In addition to music therapy, music can also be used effectively to improve the mental health of people with dementia outside of formal music therapy settings. Research has demonstrated that even where a trained music therapist is not involved, music-listening programs following evidence-based protocols can reduce agitation (Park & Specht, 2009; Park, 2010), and

Alzheimer's Disease
DOI: https://doi.org/10.1016/B978-0-12-821334-6.00009-0

anxiety. Music interventions can also improve communication, increase engagement and alertness in individuals with dementia, and enhance relationships with caregivers (Garrido, Dunne, Stevens, & Chang, 2020). This chapter will further explore what both music therapy and nontherapist-led music interventions involve and how they can improve psychological and cognitive well-being in some individuals with dementia.

Music therapy

The management of psychological symptoms such as depression, anxiety, and stress in people with dementia is complex. The presence of multiple comorbidities means that pharmacological treatment is not always ideal (Sacchetti, Turrina, & Valsecchi, 2010). Thus several decision-making bodies from around the world recommend that nonpharmacological treatments be the first-line approach to managing the symptoms of dementia (Caltagirone et al., 2005; Ouslander, Bartesl, & Beck, 2003). Thus music therapy has become a popular alternative to pharmacology for managing psychological symptoms in people with dementia such as depression, anxiety, and stress. In the following, we discuss the impact of music therapy on (1) anxiety and agitation and (2) mood and depression, respectively.

Anxiety and agitation

One of the strongest bodies of evidence for the benefit of music therapy is in relation to its capacity to reduce agitation in people with dementia. For example, Raglio et al. (2010) evaluated the effectiveness of music therapy on individuals with severe dementia with behavioral disturbances over 6 months. A total of 60 participants were split into control and experimental groups. Individuals in the experimental group participated in three cycles of 12 music therapy sessions over a month, which involved music improvisation. Throughout the study, participants were assessed using the following measures: the Mini-Mental State Examination (MMSE), Barthel Index for activities for daily living, and neuropsychiatric inventory. The music therapy sessions were nonverbal and mainly involved sound, music, and improvising to work on participants' behaviors, and emotions by expressing feelings using musical instruments. Raglio et al. (2010) found that the music therapy program improved and helped manage behavioral and psychological symptoms of dementia, particularly agitation (Raglio et al., 2010). Such findings suggest that only after a month of music therapy, individuals' behavioral disturbance symptoms improved, thus supporting the effectiveness of music therapy in individuals with dementia.

Similarly, Ridder, Stige, Qvale, and Gold (2013) explored whether music therapy can reduce behavioral and psychological symptoms of agitation in individuals with moderate to severe dementia. They were also interested in whether music therapy can improve the quality of life and impact psychotropic medication use. The music therapy sessions in the study were person-centered, meaning they were conducted differently for different individuals, and involved improvisation through singing or playing instruments, dancing, and other activities, such as talking and going for walks. A sample of 42 participants took part in the study, 21 in the music therapy group and 21 in the control or standard care group. After 6 weeks of music therapy, participants who engaged in therapy were found to have reduced agitation in comparison to individuals who did not receive music therapy. In terms of psychotropic medication usage, participants who attended music therapy did not have any significant changes to their medication use, whereas individuals who did not receive music therapy had an increase in their medication usage (Ridder et al., 2013). These findings provide further evidence that music therapy can reduce agitation as well as improve individuals' quality of life and mental health.

Sung, Lee, Li, and Watson (2012) demonstrated that group music therapy sessions were similarly found to be effective, who assessed how group music interventions affected the levels of anxiety and agitation in individuals with dementia. There were two participant groups made up of 27 participants in the experimental group and 28 participants in the control group. The experimental music intervention used in Sung et al. (2012) involved participants playing percussion instruments, such as the tambourine and handbells, as a group for 30 minutes twice a week for 6 weeks. Music and songs used in the sessions were familiar to participants and accounted for participants' music preferences. Compared to the control group, participants in the experimental group had significantly reduced anxiety after taking part in the music performance. Furthermore, participants in the experimental group reported reduced agitation, but the improvement was not significant when compared to participants in the control group. However, Sung et al. (2012) suggest this may be related to individuals not presenting agitated behavior as much as anxiety, which could have limited significant patterns of improvement or reduction of agitation in the participants (Sung et al., 2012). Moreover, participants in the control group also had somewhat reduced anxiety and agitation (Sung et al., 2012). Sung et al. (2012) suggested that this finding is related to individuals in the experimental group who had improved emotions and mood affecting individuals in the control group over time as they lived in the same residential care facility. They argue that transference plays a role in improving mood in individuals with dementia, that is, participants who had improved mood projected their positive feelings onto other individuals in their environment, consequently improving other individuals' mood.

Other studies have considered the mechanisms by which music can influence people with dementia in more detail. For example, McDermott, Orrell, and Ridder (2014) used interviews to explore individuals with dementia's perspectives on music as well as how others, such as their families, nursing staff, and music therapists, perceive the effects of music therapy. Findings revealed that music therapy allowed individuals with dementia to reflect and relate to the world, their past experiences, and was a way of improving behavioral problems (e.g., agitation) and encouraging an appreciation for life. Staff caring for individuals with dementia noted that music therapy has a relaxing effect on individuals with dementia' mood, improving their living environment. Music choice was related to their personalities, life experiences, cultural identity, and era, allowing them to have a connection to the music listened to during therapy sessions (McDermott et al., 2014). Such findings suggest that music interventions can have a psychosocial effect by allowing individuals to communicate, express feelings, decrease their agitation, and reconnect with oneself and identity.

Mood and depression

Several studies have explored the effect of music therapy on individuals with dementia who are also experiencing depressive symptoms. For example, Ashida (2000) investigated the outcomes of music therapy, as a form of reminiscence, on depression in 20 participants with dementia. Similar to other studies, such as Bevins, Dawes, Kenshole, and Gaussen (2015) and Ridder et al. (2013), the music therapy sessions involved participants playing musical instruments, such as drums, guitar, and participants singing songs they made up on the spot based on their thoughts or looking outside a window and singing about the weather. The study found that music therapy led to reduced depression and better communication skills. In addition, music therapy reduced social withdrawal and helped participants engaged more socially (Ashida, 2000), suggesting an improvement in their mood.

Another study also assessed the impact of music therapy on reducing depression. Chu et al. (2014) studied the effects of a variety of music therapy techniques on reducing depression in individuals with dementia, including choosing songs to play, listening to music, and playing instruments. Psychiatrists closely monitored participants' suicidal ideation and changes in depressive symptoms over the duration of the 6-week music therapy program. Analyses of saliva samples were also conducted to assess cortisol measures, which indicates the presence of elevated stress levels and can also be indicative of the development of depression (Pulopulos, Baeken, & De Raedt, 2020; Pulopulos, Hidalgo, Puig-Perez, Montoliu, & Salvador, 2020). Participants' cognition was also regularly measured using the MMSE. The study reported reduced depression and improved cognition, especially for individuals with mild and moderate dementia, following music therapy. However, analyses did not find any improvement in cortisol levels (Chu et al., 2014). Thus there is a strong evidence that music therapy can reduce depressive symptoms and cognition in people with dementia. It is suggested that music involves rhythm and lyrics which can stimulate cognitive functions, such as short-term memory, attention, calculation, and language (Chu et al., 2014). An interesting point to mention is that music therapy stimulates language processes, which are typically impacted by depressive symptoms (Potkins et al., 2003). Importantly, music is not as effective in individuals with severe dementia (Garrido, Dunne, Perz, Chang, & Stevens, 2018; Garrido, Stevens, Chang, Dunne, & Perz, 2018). This is perhaps because severe dementia is associated with more widespread brain damage and with several cognitive problems. Widespread disease manifestation limits the ability or potential for music to stimulate cognitive functions, such as memory, which could be diminished in the later stages of dementia.

Some other studies investigated the impact of music therapy on mood in general. Bevins et al. (2015) evaluated the effectiveness of music therapy in individuals with dementia who also had intellectual disabilities, such as Down syndrome. Participants took part in 3 assessments and then 18 intervention sessions of music therapy. The music therapy sessions generally involved participants sharing their feelings through music, playing musical instruments and engaging in musical intercommunication through play. Participants and support workers took part in semistructured interviews. The study reported that the music therapy group found sessions pleasurable and exciting to attend. Moreover, as a result of the music therapy sessions, most participants were also reported to have improved communication and mood (Bevins et al., 2015).

Nontherapist-led music interventions

In addition to music therapy, music interventions that are not led by a music therapist can also have positive impacts on psychological well-being in people with dementia. Such interventions range from the use of background music in aged care facilities to the use of personalized playlists using standardized protocols and live music events in care homes.

Well-being, anxiety, mood

Several studies have investigated the impact of music interventions in mental health in dementia patients. For example, van der Vleuten, Visser, and Meeuwesen (2012) explored the effect of live music performance on individuals with dementia in terms of well-being and the quality of life. In a sample of 54 individuals with mild and severe dementia, most participants enjoyed live music performances and experienced positive effects on of viewing performances on their well-being. Results suggested that live music performances may have a more lasting positive impact on individuals with mild dementia, while for individuals with severe dementia, it may be an in-the-moment type of joy or pleasure. This finding is in agreement with other studies showing the impact of music therapy on improving well-being is related to disease stage and symptom severity in dementia patients (Garrido, Dunne, et al., 2018; Garrido, Stevens, et al., 2018).

Since there is an evidence suggesting music preference influences the experience and effectiveness of music on well-being (McDermott et al., 2014), Sung, Chang, and Lee (2010) explored how listening to preferred music can affect anxiety in individuals with dementia. They assigned 29 individuals to the experimental group, while 23 were assigned to the control group to compare listening to preferred music. Both participant groups received standard care, such as participating in social activities and exercise, but participants in the experimental group also took part in music-listening sessions twice a week and for 6 weeks. The Music Preference Scale was used to explore participants' music preferences, and the Rating Anxiety in Dementia Scale was used to measure participants' signs and symptoms of anxiety. In line with McDermott et al. (2014), Sung et al. (2010) found that listening to preferred music significantly reduced participants' anxiety. Sung et al. (2010) suggest that the effectiveness of preferred music in reducing anxiety symptoms is related to individuals' familiarity with the music, such as the memories and the environment it imposes as eliciting positive emotions for individuals with dementia. Findings of studies highlight the importance of taking into consideration individuals' music preferences before implementing music interventions for individuals with dementia.

Negative versus positive impact of music interventions

While popular opinions can tend to portray music as a "cure all" for people with dementia, research indicates that music interventions may either have a positive or negative impact on individuals with dementia. For example, Ziv, Granot, Hai, Dassa, and Haimov (2007) explored the effects of background music on positive and negative behaviors in individuals with Alzheimer's disease (AD). In this study, 28 participants with moderate AD listened to happy, upbeat music that was familiar to them. Researchers observed and categorized participants' behaviors as positive, negative, or neutral. Positive behavior was displayed by being talkative, laughing, smiling, calming others, and singing. Negative behavior included expressing agitation, aggression, complaining, repetitive behaviors, crying, and harassing others. Neutral behavior involved sleeping and staring into space. Music sessions took place once a week for 3 weeks. The results of the study indicated that when participants listened to music, they showed increased positive behavior and reduced negative behavior. In nine participants, negative behavior ceased, regardless of whether they were listening to background music or not. However, out of the 28 participants, nine participants had no response, and two participants showed negative behavior when listening to background music (Ziv et al., 2007).

Similarly, in a personalized playlist intervention following an evidence-based protocol, Garrido et al. (2020) found that positive behaviors increased in the majority of listening sessions with participants. However, mood deterioration was reported in 11 listening situations involving 4 different participants. These findings highlight the importance of identifying people with dementia who may be vulnerable to negative responses due to their mental health history, selecting music for those individuals with additional care, and monitoring responses (Chu et al., 2014; Garrido et al., 2020).

Despite the potential for negative responses, research suggests that residential aged care facilities do not always recognize the need for caution or have the required knowledge to implement music interventions effectively. To investigate the implementation of music interventions in aged care settings, Garrido, Dunne, et al. (2018) and Garrido, Stevens, et al. (2018) conducted an online questionnaire and interviews with staff across multiple residential aged facilities. Results indicated that most aged care facilities incorporated music interventions in their setting, but they were not music interventions that met evidence-based standards. Music programs seldom involved a professional music therapist, and although opinions about music interventions were primarily positive, cases of individuals who responded negatively were reported (Garrido, Dunne, et al., 2018; Garrido, Stevens, et al., 2018). Negative responses could occur when the music was too loud or repetitive, or when the music elicited distressing memories (Garrido, Dunne, et al., 2018; Garrido, Stevens, et al., 2018). However, limitations to available resources, staff buy-in, and know-how were identified as significant barriers to the effective implementation of music interventions. In fact, staff knowledge can have a considerable influence on the effectiveness of music interventions. Sung, Lee, Chang, and Smith (2011) explored 214 nurses' attitudes and expectations of music for individuals with dementia. After conducting questionnaires, nurses had positive attitudes toward the use of music and music therapy for individuals with dementia in nursing homes. As in the Garrido, Dunne, et al. (2018) and Garrido, Stevens, et al. (2018) studies, it was found that nursing staff did not have much training and education on the use or implementation of music (Sung et al., 2011). The authors suggest that this should be a part of training for nursing staff in aged care settings.

Theories of how music therapies and interventions work

It is important to note that the exact mechanism of how music therapy and other music interventions work is not entirely clear, although there are several theories. Some argue that social processes are some of the factors underlying the effectiveness of music therapy and music interventions in people with dementia. Like music therapy, research has identified how social processes can influence responses to nontherapist-led interventions. Sherratt, Thornton, and Hatton (2004) explored how social interactions, as a component of live music, affect participants' behavior, in comparison to prerecorded music. Researchers noted and observed behavioral indicators of participants' engagement in four conditions: no music, commercially recorded and taped music, a taped version of a musician singing, and a live performance of music sung by the same musician. Song selections were the same across the four conditions. Results showed that participants showed the greatest engagement in the live music condition. Such findings suggest that incorporating opportunities for social interaction in musical experiences can enhance the well-being of dementia patients. In addition, a factor that is a critical part of music therapy, music, can regulate the levels of arousal and increase engagement with other people in the environment (Ridder, 2011).

In addition to social processes related to music, it is also argued that music therapy and music interventions also engage the limbic system (Moore, 2013), which plays a key role in emotions, such as fear and stress (Rajmohan & Mohandas, 2007; Rolls, 2015, 2019). It is possible then by engaging the limbic system, music therapy decreases stress and anxiety, as reported in several prior studies (Moore, 2013). Another theory argued that music regulates arousal levels and can then subsequently impact cognitive functioning (Schafer, Sedlmeier, Stadtler, & Huron, 2013).

Conclusions and future studies

The studies reviewed in this chapter have demonstrated that music therapy and nontherapist-led interventions have great potential to help alleviate psychological symptoms such as agitation, anxiety, and depression in people with dementia. The review has highlighted the importance of using music that is of personal relevance to individuals with dementia as well as to be aware of the potential for negative responses. Professional music therapists are trained both in the personalization of music treatments and in managing negative responses. However, where music interventions are conducted outside of a formal music therapy context, there is a need for care staff to receive further training in order to implement music programs most effectively.

Fortunately, evidence-based guidelines for developing music interventions in aged care settings are available (Clements-Cortes, Pearson, & Chang, 2015; Garrido, Dunne, Stevens, Chang, & Perz, 2019), and research has demonstrated that where such protocols are followed, outcomes are more positive than a more ad hoc approach (Garrido et al., 2017). Nevertheless, ongoing challenges to the implementation of music interventions in aged care settings still exist. Aged care staff are typically overburdened and have little time to undertake additional training or to implement new programs in their care environments. Furthermore, there is often a culture in aged care facilities in which emotional or psychological needs are deemed secondary to physical ones. Both government policy and market forces are pushing toward "whole-person" and "person-centered" approaches to caring for people with dementia (Manthorpe & Samsi, 2016). Music interventions can play an important role in these models of care, since music is something that reflects our identity as individuals and can reconnect people with dementia with their sense of self (Evans, Garabedian, & Bray, 2017). Future research will need to consider how music and music therapy can become further embedded in practices of care for people with dementia.

References

Ashida, S. (2000). The effect of reminiscence music therapy sessions on changes in depressive symptoms in elderly persons with dementia. *Journal of Music Therapy*, *37*(3), 170–182. Available from https://doi.org/10.1093/jmt/37.3.170.

Bevins, S., Dawes, S., Kenshole, A., & Gaussen, K. (2015). Staff views of a music therapy group for people with intellectual disabilities and dementia: A pilot study. *Advances in Mental Health and Intellectual Disabilities*, *9*(1), 40–48. Available from https://doi.org/10.1108/AMHID-04-2014-0005.

Caltagirone, C., Bianchetti, A., Di Luca, M., Mecocci, P., Padovani, A., Pirfo, E., & Italian Association of Psychogeriatrics. (2005). Guidelines for the treatment of Alzheimer's disease from the Italian Association of Psychogeriatrics. *Drugs & Aging*, *22*(Suppl. 1), 1–26.

Chu, H., Yang, C. Y., Lin, Y., Ou, K. L., Lee, T. Y., O'Brien, A. P., & Chou, K. R. (2014). The impact of group music therapy on depression and cognition in elderly persons with dementia: A randomized controlled study. *Biological Research for Nursing*, *16*(2), 209–217. Available from https://doi.org/10.1177/1099800413485410.

Clark, I. N., Stretton-Smith, P. A., Baker, F. A., Lee, Y. C., & Tamplin, J. (2020). It's feasible to write a song": A feasibility study examining group therapeutic songwriting for people living with dementia and their family caregivers. *Frontiers in Psychology*, *11*, 1951. Available from https://doi.org/10.3389/fpsyg.2020.01951.

Clements-Cortes, A., Pearson, C., & Chang, K. (2015). *Creating effective music listening opportunities*. Toronto: Baycrest. Retrieved from www.baycrest.org/care/culture-arts-innovation/therapeutic-arts/music-therapy/creating-effective-music-listeningopportunities.

Davidson, J., & Garrido, S. (2015). Singing and psychological needs. In G. Welsh, D. Howard, & J. Nix (Eds.), The Oxford handbook of singing. Oxford: Oxford University Press.

Evans, S. C., Garabedian, C., & Bray, J. (2017). 'Now he sings'. The My Musical Memories Reminscence Programme: Personalised interaction reminiscence sessions for people living with dementia. *Dementia (Basel, Switzerland)*, *18*(3), 1181–1198.

Garrido, S., Dunne, L., Chang, E., Perz, J., Stevens, C., & Haertsch, M. (2017). The use of music playlists for people with dementia: A critical synthesis. *Journal of Alzheimer's Disease*, *60*, 1129–1142.

Garrido, S., Dunne, L., Perz, J., Chang, E., & Stevens, C. J. (2018). The use of music in aged care facilities: A mixed-methods study. *Journal of Health Psychology*, *25*(10–11), 1425–1438. Available from https://doi.org/10.1177/1359105318758861.

Garrido, S., Dunne, L., Stevens, C., & Chang, E. (2020). Music playlists for people with dementia: Trialing a guide for caregivers. *Journal of Alzheimer's Disease*, 77(1), 219–226.

Garrido, S., Dunne, L., Stevens, C., Chang, E., & Perz, J. (2019). *Music playlists for people with dementia: A guide for carers, health workers and family*. Retrieved from https://musicfordementia.github.io/.

Garrido, S., Stevens, C., Chang, E., Dunne, L., & Perz, J. (2018). Music and dementia: Individual differences in response to personalized playlists. *Journal of Alzheimer's Disease*, *64*(3), 933–941. Available from https://doi.org/10.3233/JAD-180084.

Lam, H. L., Li, W. T. V., Laher, I., & Wong, R. Y. (2020). Effects of music therapy on patients with dementia – A systematic review. *Geriatrics (Basel)*, *5*(4), 62. Available from https://doi.org/10.3390/geriatrics5040062.

Leggieri, M., Thaut, M. H., Fornazzari, L., Schweizer, T. A., Barfett, J., Munoz, D. G., & Fischer, C. E. (2019). Music intervention approaches for Alzheimer's disease: A review of the literature. *Frontiers in Neuroscience*, *13*, 132. Available from https://doi.org/10.3389/fnins.2019.00132.

Manthorpe, J., & Samsi, K. (2016). Person-centered dementia care: Current perspectives. *Clinical Interventions in Aging*, *11*, 1733–1740. Available from https://doi.org/10.2147/CIA.S104618.

McDermott, O., Orrell, M., & Ridder, H. M. (2014). The importance of music for people with dementia: The perspectives of people with dementia, family carers, staff and music therapists. *Aging & Mental Health*, *18*(6), 706–716. Available from https://doi.org/10.1080/13607863.2013.875124.

Moore, K. S. (2013). A systematic review on the neural effects of music on emotion regulation: Implications for music therapy practice. *Journal of Music Therapy*, *50*(3), 198–242.

Ouslander, J., Bartesl, S., & Beck, C. (2003). Consensus statement on improving the quality of mental health care in US nursing homes: Management of depression and behavioral symptoms associated with dementia. *Journal of the American Geriatric Society*, *51*, 1287–1298.

Park, H. (2010). Effect of music on pain for home-dwelling persons with dementia. *Pain Management Nursing*, *11*, 141–147.

Park, H., & Specht, J. (2009). Effect of individualized music on agitation in individuals with dementia who live at home. *Journal of Gerontological Nursing*, *35*, 47–55.

Perez-Eizaguirre, M., & Vergara-Moragues, E. (2020). Music therapy interventions in palliative care: A systematic review. *Journal of Palliative Care*. available from https://doi.org/10.1177/0825859720957803.

Potkins, D., Myint, P., Bannister, C., Tadros, G., Chithramohan, R., Swann, A., & Margallo-Lana, M. (2003). Language impairment in dementia: Impact on symptoms and care needs in residential homes. *International Journal of Geriatric Psychiatry*, *18*(11), 1002–1006. Available from https://doi.org/10.1002/gps.1002.

Pulopulos, M. M., Baeken, C., & De Raedt, R. (2020). Cortisol response to stress: The role of expectancy and anticipatory stress regulation. *Hormones and Behavior*, *117*, 104587. Available from https://doi.org/10.1016/j.yhbeh.2019.104587.

Pulopulos, M. M., Hidalgo, V., Puig-Perez, S., Montoliu, T., & Salvador, A. (2020). Relationship between cortisol changes during the night and subjective and objective sleep quality in healthy older people. *International Journal of Environmental Research and Public Health*, *17*(4), 1264. Available from https://doi.org/10.3390/ijerph17041264.

Raglio, A., Bellelli, G., Traficante, D., Gianotti, M., Ubezio, M. C., Gentile, S., & Trabucchi, M. (2010). Efficacy of music therapy treatment based on cycles of sessions: A randomised controlled trial. *Aging and Mental Health*, *14*(8), 900–904. Available from https://doi.org/10.1080/13607861003713158.

Rajmohan, V., & Mohandas, E. (2007). The limbic system. *Indian Journal of Psychiatry*, *49*(2), 132–139. Available from https://doi.org/10.4103/0019-5545.33264.

Ridder, H. M. O. (2011). How can singing in music therapy influence social engagement for people with dementia? Insights from the polyvagal theory. In F. Baker, & S. Uhlig (Eds.), *Voicework in music therapy: Research and practice* (pp. 130–146). London: Jessica Kingsley Publishers.

Ridder, H. M. O., Stige, B., Qvale, L. G., & Gold, C. (2013). Individual music therapy for agitation in dementia: An exploratory randomized controlled trial. *Aging & Mental Health, 17*(6), 667–678. Available from https://doi.org/10.1080/13607863.2013.790926.

Rolls, E. T. (2015). Limbic systems for emotion and for memory, but no single limbic system. *Cortex; A Journal Devoted to the Study of the Nervous System and Behavior, 62*, 119–157. Available from https://doi.org/10.1016/j.cortex.2013.12.005.

Rolls, E. T. (2019). The cingulate cortex and limbic systems for action, emotion, and memory. *Handbook of Clinical Neurology, 166*, 23–37. Available from https://doi.org/10.1016/B978-0-444-64196-0.00002-9.

Sacchetti, E., Turrina, C., & Valsecchi, P. (2010). Cerebrovascular accidents in elderly people treated with antipsychotic drugs. *Drug Safety, 33*, 273–288.

Schafer, T., Sedlmeier, P., Stadtler, C., & Huron, D. (2013). The psychological functions of music listening. *Frontiers in Psychology, 4*, 511. Available from https://doi.org/10.3389/fpsyg.2013.00511.

Sherratt, K., Thornton, A., & Hatton, C. (2004). Emotional and behavioural responses to music in people with dementia: An observational study. *Aging & Mental Health, 8*(3), 233–241. Available from https://doi.org/10.1080/13607860410001669769.

Sung, H. C., Chang, A. M., & Lee, W. L. (2010). A preferred music listening intervention to reduce anxiety in older adults with dementia in nursing homes. *Journal of Clinical Nursing, 19*(7–8), 1056–1064. Available from https://doi.org/10.1111/j.1365-2702.2009.03016.x.

Sung, H. C., Lee, W. L., Chang, S. M., & Smith, G. D. (2011). Exploring nursing staff's attitudes and use of music for older people with dementia in long-term care facilities. *Journal of Clinical Nursing, 20*(11–12), 1776–1783. Available from https://doi.org/10.1111/j.1365-2702.2010.03633.x.

Sung, H. C., Lee, W. L., Li, T. L., & Watson, R. (2012). A group music intervention using percussion instruments with familiar music to reduce anxiety and agitation of institutionalized older adults with dementia. *International Journal of Geriatric Psychiatry, 27*(6), 621–627. Available from https://doi.org/10.1002/gps.2761support/caregiving/stages-behaviors/depression.

van der Vleuten, M., Visser, A., & Meeuwesen, L. (2012). The contribution of intimate live music performances to the quality of life for persons with dementia. *Patient Education and Counseling, 89*(3), 484–488. Available from https://doi.org/10.1016/j.pec.2012.05.012.

Ziv, N., Granot, A., Hai, S., Dassa, A., & Haimov, I. (2007). The effect of background stimulative music on behavior in Alzheimer's patients. *Journal of Music Therapy, 44*(4), 329–343. Available from https://doi.org/10.1093/jmt/44.4.329.

CHAPTER 12

The efficacy of donepezil for the treatment of Alzheimer's disease

Samuel L. Warren[1] and Ahmed A. Moustafa[2,3,4]
[1]Psychological Science, School of Psychology, Western Sydney University, Sydney, NSW, Australia
[2]School of Psychology, Western Sydney University, Sydney, NSW, Australia
[3]MARCS Institute for Brain and Behaviour, Western Sydney University, Sydney, NSW, Australia
[4]Department of Human Anatomy and Physiology, The Faculty of Health Sciences, University of Johannesburg, Johannesburg, South Africa

Introduction

Alzheimer's disease (AD) is a neurodegenerative disorder that involves several pathological changes such as amyloid plaque accumulation, cognitive decline, neurofibrillary tangles, and secondary inflammation (Alzheimer's Association, 2018). In AD, these abnormal neurological symptoms are incredibly severe and progressively worsen to the point of death. Consequentially, there is a critical need for interventions that can treat the progressive symptoms of AD. The most common type of drugs used to treat the symptoms of AD is cholinesterase inhibitors (Singh & Sadiq, 2020). Cholinesterase inhibitors work by regulating the neurotransmitter acetylcholine, which is often depleted during AD. By maintaining cholinergic function, cholinesterase inhibitors are thought to be protective against, and slow, some of the symptoms of AD. Accordingly, the literature suggests that the administration of cholinesterase inhibitors could treat some of the progressive symptoms of AD (e.g., cognitive decline) (Zemek et al., 2014; Zhang, Yu, Wang, & Zheng, 2020).

The drugs donepezil, rivastigmine, and galantamine are the most common forms of cholinesterase inhibitors used to treat AD (Sharma, 2019). Each of these cholinesterase inhibitors has varying therapeutic effects; however, each drug is predominantly used to treat the neuropsychological symptoms of AD (e.g., memory, attention). In turn, some studies have suggested that cholinesterase inhibitors can improve cognition and daily function in AD (Espiritu & Cenina, 2020; Ma, Ji, Li, Yang, & Pan, 2018). For example, donepezil, galantamine, and rivastigmine have all been observed to improve memory function in AD (Haake, Nguyen, Friedman, Chakkamparambil, & Grossberg, 2020).

Alzheimer's Disease
DOI: https://doi.org/10.1016/B978-0-12-821334-6.00001-6

There is also an evidence to suggest that cholinesterase inhibitors can affect the neurodegenerative symptoms of AD. For example, some studies have shown that cholinesterase inhibitors can help maintain the structure and function of the hippocampus in AD (Moss, 2020). However, this neurodegenerative research is highly debated. Accordingly, there are some contentions concerning the specific effects and uses of cholinesterase inhibitors.

While cholinesterase inhibitors have some benefits, the effectiveness of these drugs depends on individual factors such as participant age, disease stage (mild vs advanced), and drug dosage. The effects of cholinesterase inhibitors can also be diminished by adverse events, which can deter some individuals from undergoing treatment. For example, some individuals can experience adverse events such as nausea, vomiting, and diarrhea when taking cholinesterase inhibitors (Doody et al., 2012). Consequentially, there is a contention concerning the efficacy and safety of cholinesterase inhibitors. There is also a contention about how best to use cholinesterase inhibitors for the treatment of AD. In this chapter, we discuss the efficacy and safety of using donepezil for the treatment of AD. Specifically, we discuss the following: (1) The therapeutic effects of donepezil, (2) the adverse effects of donepezil, (3) directions for future donepezil research, and (4) combined donepezil treatments [donepezil and memantine or cognitive stimulation therapy (CST)]. In turn, we aim to provide an understanding of the effects of donepezil in AD and highlight areas for future research.

It is important to note that, in this chapter, we exclusively discuss donepezil over other cholinesterase inhibitors. We specifically focus on donepezil because (1) it is the most popular cholinesterase inhibitor, (2) it is the only drug that has been approved for the treatment of all stages of AD, and (3) it has additional benefits compared to other cholinesterase inhibitors (e.g., rivastigmine and galantamine) (Alzheimer's Association, 2020; Kobayashi, Ohnishi, Nakagawa, & Yoshizawa, 2016). These benefits include improved neuropsychological effects, dosage and administration effects (e.g., Its oral form is easily ingested only once a day unlike rivastigmine), and its debated cost-effectiveness (Benjamin & Burns, 2007; Blanco-Silvente et al., 2017).

The effects of donepezil in Alzheimer's disease

The effects of donepezil on the neuropsychological symptoms of AD

In AD research, the development and use of cholinesterase inhibitors are largely based on the cholinergic hypothesis. The cholinergic hypothesis originally stated that the depletion and dysfunction of the neurotransmitter acetylcholine causes cognitive decline in AD (Bartus, Dean, Beer, & Lippa, 1982). However, modern iterations of the theory consider cholinergic decline to be a risk factor or contributory process toward cognitive decline in AD (Craig, Hong, & McDonald, 2011). The cholinergic system of the brain is instrumental in managing attention, neuroplasticity, and complex cognitive processes such as memory (Teipel, Grinberg, Hampel, & Heinsen, 2009). Accordingly, there is a clear link between the higher-order symptoms of AD and tasks that involve the cholinergic system. For example, neurodegeneration in the basal forebrain is related to the loss of acetylcholinic neurons and memory function in AD (Ferreira-Vieira, Guimaraes, Silva, & Ribeiro, 2016; Francis, Palmer, Snape, & Wilcock, 1999). In healthy aging, it is typical for cholinergic neurons to degenerate slowly (Davidson & Winocur, 2010); however, modern iterations of the cholinergic hypothesis state that in AD, cholinergic decline is drastically increased by senile plaques and neurofibrillary tangles (Craig et al., 2011). Accordingly, contemporary theories such as the amyloid cascade hypothesis regard cholinergic decline as a symptom of AD (not a cause as once thought) (Armstrong, 2013). Nonetheless, to date, cholinesterase inhibitors are the only medications that viably treat and delay symptoms of AD (apart from memantine).

In the literature, it is widely agreed that donepezil can treat the neuropsychological symptoms of AD (Cacabelos, 2007; Haake et al., 2020). For example, in a study by Chang, Yang, Chou, Chen, and Yang (2015), donepezil was found to positively affect the symptoms of cognitive and functional decline in AD. These results were found by comparing the Mini-Mental State Examination (MMSE), Cognitive Abilities Screening Instrument, and Clinical Dementia Rating (CDR) Scale scores of donepezil-treated and control participants. Chang et al. (2015) also found that MMSE and CDR scores were related to participants withdrawal from treatment. Specifically, they stated that an increase of 1 unit on the MMSE resulted in a decreased chance of treatment dropout (6%), while a one-point rise in CDR scores increased the chance of treatment withdrawal (10.5%). Accordingly, Chang et al. (2015) suggested that early donepezil treatment can slow down the initial symptoms of cognitive and functional decline in AD and could be protective against treatment dropout.

It has also been observed that increases in donepezil dosage (from 5 to 10 mg or 23 mg) can further improve the neuropsychological symptoms of AD. For example, a study by Farlow et al. (2010) compared the safety and efficacy of 10 and 23 mg doses of donepezil when treating AD. The different dosages were compared to understand whether a higher donepezil dosage could improve the symptoms of AD more so than a standard 10 mg dose. Approximately 1451 participants were recruited and split into the two 23-mg experimental ($n = 972$) and 10-mg standard dosage ($n = 479$) groups. Cognitive and psychological measures were recorded using the Severe impairment battery (SIB)-cognitive, MMSE, Activities of Daily Living (ADL) Scale, and the Clinician's Interview-Based Impression of Change Plus Caregiver Input Scale. Farlow et al. (2010) found that participants in the 23-mg donepezil dosage group had better cognitive scores than individuals in the 10-mg group (as measured by the SIB). They also suggested that a higher dose of donepezil was more effective in individuals with late-stage AD. Accordingly, Farlow et al. (2010) concluded that donepezil can treat the cognitive symptoms of AD and that increased dosages (23 mg) can further improve these symptoms. Correspondingly, similar dosage effects have also been observed between 5 and 10 mg of donepezil in the wider literature (Birks & Harvey, 2006; Yatabe et al., 2013).

However, it is important to note that some studies have found that donepezil does not affect cognitive and functional ability. For example, Andersen et al. (2012) studied the effects of donepezil on the neuropsychological symptoms of AD and received negative results. Specifically, Andersen et al. (2012) found no significant differences in cognitive scores between participants treated with donepezil or a placebo. However, it is important to note that participants experienced significant side effects including anorexia and gastrointestinal issues leading to participant dropouts (25% in the drug group). In turn, potential cognitive improvement in the study by Andersen et al. (2012) may have been obscured by the severe adverse effects of donepezil. In the wider literature, evidence suggests that—while donepezil can improve cognitive and functional ability—treatment success is reliant on the dosage of donepezil, physiology of the individual, and the stage of AD. Accordingly, the negative results of some donepezil studies, such as Andersen et al. (2012), may have occurred because the adverse side effects of donepezil outweighed the benefits of the drug in the study population.

The adverse effects of donepezil

Clearly, donepezil has the capacity to treat some of the neuropsychological symptoms of AD. However, there is a relationship between drug dosage, treatment efficacy, and the adverse effects of donepezil that can limit treatment capacity. Most, if not all, of the studies noted previously that report positive effects for donepezil also report numerous adverse side effects. Accordingly, it is important to ask whether the therapeutic benefits outweigh the adverse effects of donepezil. Moreover, it is important to understand how changes in drug dosage and environmental factors influence treatment outcomes.

It is common for donepezil to have adverse effects that lead to participant discomfort or dropout. For example, in a study by Griffith, Lichtenberg, Goldman, and Payne-Parrish (2006), 64 participants experienced 151 adverse events due to donepezil. Specifically, participants experienced side effects such as diarrhea, hypertension, urinary tract infection, headaches, dizziness, anorexia, and nausea. Because of these adverse events, 16 participants either reduced their donepezil dosage from 10 to 5 mg or dropped out of the study entirely. In a review on the topic, Adlimoghaddam, Neuendorff, Roy, and Albensi (2018) found that donepezil-treated participants had improved cognitive and functional symptoms; however, they also reported that donepezil treatment resulted in an increased participant dropout rate (especially at higher drug dosages). Accordingly, some participants may be missing treatment due to the adverse effects of donepezil.

However, it is important to note that, while the adverse effects of donepezil occur frequently in research, they occur in a minority of participants. Moreover, studies suggest that the adverse effects of donepezil do not often outweigh the therapeutic effects of the drug (Jackson, Ham, & Wilkinson, 2004). For example, a study by Jia et al. (2017) investigated the safety of donepezil treatment in participants with mild to moderate AD. Specifically, Jia et al. (2017) conducted an in-depth double-blind study on the adverse events that occur when taking 10 mg of donepezil. Approximately 241 participants were administered 5 mg of donepezil a day for 4 weeks followed by 10 mg for 20 weeks. The MMSE, ADL Scale, *APOE4* gene, and electrocardiograms were used to study participants response to donepezil. Jia et al. (2017) found that 93 out of the initial 241 participants reported adverse events; however, none of these adverse events was serious to the point of hospitalization or causing significant harm. It was also noted that participants were found to have better cognitive scores (MMSE) when taking a 10-mg dosage of donepezil (compared to 5 mg). Moreover, it was observed that participants with the *APOE4* gene or cardiovascular disorders had an increased chance of adverse events occurring. Consequentially, Jia et al. (2017) concluded that a 10-mg dose of donepezil is safe and that individual factors (e.g., cardiovascular health, *APOE4*) affect adverse events more so than dosage.

Accordingly, current evidence suggests that the therapeutic effects of donepezil supersede the adverse effects of the drug. However, although the 10- and 23-mg dosages can decrease the symptoms of cognitive and functional decline in AD, the efficacy of these dosages is dependent on the admiration of the drug and the individual. Even when some symptoms of AD are improved, there are still adverse events that can diminish donepezil treatment. Subsequently, researchers and clinicians should take precautions when treating AD with donepezil. The literature suggests that participants respond to 10-mg dosages of donepezil best when gradually transitioned to such from a 5-mg dose. Nonetheless, it is clear from the literature that a 10- or 23-mg dosage is not always the best fit for an individual and may result in treatment dropout or a lower quality of life (QOL) (Rogers, Farlow, Doody, Mohs, & Friedhoff, 1998). Accordingly, it is imperative that researchers observe individual's reaction to donepezil and adjust the treatment accordingly. We recommend a 5-mg dose of donepezil as it is considerably safer than higher doses (10- or 23-mg dose) and can still significantly improve cognition, daily function, and QOL in AD.

The potential effects of donepezil in AD

As discussed previously, cholinesterase inhibitor research predominantly focuses on treating the neuropsychological symptoms of AD; however, some preliminary research has also suggested that donepezil can treat other symptoms of AD. For example, research has suggested that donepezil could be used to treat some of the neurodegenerative, metabolic, and neuropsychiatric symptoms of AD (Cummings et al., 2016; Dubois et al., 2015; Moon, Kim, & Jeong, 2016). However, it is important to note that the treatment of these symptoms is heavily debated. In this section, we explore some of the potential effects of donepezil therapy in AD. Accordingly, we aim to highlight the emerging areas of donepezil research and inspire further investigation.

The cholinergic hypothesis states that the function of the cholinergic system is closely tied to neurodegenerative mechanisms in AD. Accordingly, some studies have suggested that donepezil could slow neurodegeneration in AD (Dubois et al., 2015; Moss, 2020). For example, a study by Ishiwata, Mizumura, Mishina, Yamazaki, and Katayama (2014) found that donepezil therapeutically affected symptoms of neurodegeneration in AD. Specifically,

Ishiwata et al. (2014) assessed hippocampal atrophy in 265 participants (80 controls and 185 probable AD participants) over time. Hippocampal atrophy and neuropsychological scores were measured using magnetic resonance imaging and extensive cognitive batteries. From the data collected, Ishiwata et al. (2014) conclude that there is a relationship between cognition and hippocampal atrophy in AD that is affected by donepezil. Moreover, they found that donepezil slowed hippocampal atrophy compared to controls. Accordingly, there is an evidence to suggest that donepezil can slow neurodegeneration in AD; however, it is important to note that some studies have reported contrary evidence (Wang et al., 2010). Consequentially, the relationship between donepezil and neurodegeneration needs further research.

The neurodegenerative symptoms of AD are also closely tied to metabolic abnormalities in AD. Subsequently, some preliminary research has suggested that donepezil might have therapeutic effects on brain metabolism in AD. For example, Moon et al. (2016) investigated the effects of donepezil treatment on symptoms of atrophy and metabolic decline in AD. Approximately 21 participants were studied using magnetic resonance spectroscopy and voxel-based morphometry techniques. Participants also underwent psychological testing using the Korean MMSE, Alzheimer's Disease Assessment Scale-Cognitive, CDR, and Geriatric Depression Scale. When concerning neurodegeneration, Moon et al. (2016) found that donepezil treatment increased cognitive ability and hippocampal volume. Conversely, participants who did not receive donepezil had worse cognitive scores and more hippocampal atrophy. These results were similar to Ishiwata et al. (2014) discussed previously. Moon et al. (2016) also found that donepezil treatment was protective against metabolic abnormalities. Specifically, they found that donepezil treatment resulted in an increased *N*-acetylaspartate/creatine ratio as well as a decrease in Myo-inositol/creatine and choline/creatine ratios. These metabolic ratios are important as they indicated that donepezil is protective against symptoms of amyloid/tauopathy (correlated with *N*-acetylaspartate), neurodegeneration, and cholinergic decline (choline/creatine ratio) in AD. Accordingly, Moon et al. (2016) suggested that donepezil could be used to treat neurodegeneration and metabolic abnormalities in AD.

Lastly, some preliminary research has indicated that donepezil may improve some of the neuropsychiatric symptoms of AD (e.g., depression, apathy, anxiety). For example, Cummings et al. (2016) reviewed the therapeutic effects of donepezil on the neuropsychiatric symptoms of AD. They found that donepezil improved some of the neuropsychiatric symptoms of AD such as delusions, anxiety, agitation, and irritability. These findings were dictated by differences in Neuropsychiatric Inventory (NPI) scores between donepezil and control groups. It was also found that donepezil continued to improve neuropsychiatric symptoms during late-stage AD and in clinical settings. Accordingly, Cummings et al. (2016) suggested that donepezil can improve the neuropsychiatric symptoms of AD and might even be a safe alternative to current neuropsychiatric medications for dementia. Yet, Cummings et al. (2016) did note that there is a need for great research before donepezil could be considered as a standalone treatment for the neuropsychiatric symptoms of AD. Moreover, some other studies have found that donepezil does not affect the neuropsychiatric symptoms of AD (Blanco-Silvente et al., 2017; Kobayashi et al., 2016). Consequentially, there is a need for further research into the therapeutic effects of donepezil on the neuropsychiatric symptoms of AD.

In conclusion, there are some evidences to suggest that donepezil could treat the neurodegenerative, metabolic, and neuropsychiatric symptoms of AD; however, there is also a need for further research and replication. There is also some other potential effects of donepezil that were not discussed in this chapter that require further research. For example, there is a small amount of research on the relationships between donepezil and AD biomarkers (e.g., amyloid-beta) that needs further investigation (Lu et al., 2015; Ma et al., 2018). Future research should seek to study these topics and understand the ability, efficacy, and safety of donepezil when treating these symptoms.

Combined treatments involving donepezil

While donepezil treatment can improve some of the symptoms of AD, it is clear that cholinergic inhibitors are not an all-encompassing treatment. Accordingly, research often seeks to supplement donepezil with other interventions to improve AD treatment. Two of the most common treatments combined with donepezil are CST and the drug memantine. In this section, we explore the efficacy and safety of combined treatments and compare them to standard donepezil treatment.

Combined cognitive stimulation therapy and donepezil treatment

CST is a psychological intervention that seeks to maintain and enhance the social, cognitive, and functional skills of individuals with AD

(Kim et al., 2017). Participants undergoing CST will commonly complete themed tasks and activities as part of a group workshop. These group tasks are thought to be beneficial as they promote and maintain the use of social and psychological skills in AD. For example, some studies suggest that CST can significantly improve the social skills, daily function, and QOL of individuals with AD (Cammisuli, Danti, Bosinelli, & Cipriani, 2016). Accordingly, organizations such as the UK's National Institute for Health and Care Excellence recommend CST when treating dementia (Pink, O'Brien, Robinson, & Longson, 2018).

CST is commonly combined with donepezil treatment as both interventions are thought to improve cognition and daily function. Moreover, CST is also thought to increase the QOL of individuals with AD. In turn, evidence suggests that the combination of donepezil and CST can treat AD better than each intervention can individually (Chen et al., 2019). For example, a study by Matsuda (2007) compared the effectiveness of donepezil to a combined drug and CST treatment. Both groups received a 5-mg dose of donepezil and the CST group underwent cognitive, conversational, memory, and social tasks. Matsuda (2007) found that participants in the donepezil group experienced cognitive decline while the combined therapy group did not. Accordingly, Matsuda (2007) found that, although donepezil treatment does slow cognitive decline in AD, a combined treatment is more effective.

In another study, Chapman, Weiner, Rackley, Hynan, and Zientz (2004) also found that a combined CST and donepezil treatment was more effective than donepezil alone. Specifically, Chapman et al. (2004) found that a combined treatment improved participants' cognition, daily function, emotions (e.g., reduced irritability), QOL, and verbal communication. These measures were quantified using the MMSE, NPI, QOL, and other scales. Chapman et al. (2004) concluded that CST could successfully supplement donepezil treatment resulting in the slowing of AD progression and an improvement in emotional well-being. Chapman et al. (2004) theorized that improvements occurred because the individuals undergoing CST were consistently engaged in psychological tasks that were protective against decline. However, they did also note that the effects of both treatments were delayed and took time to appear (approximately 8 months).

In conclusion, evidence suggests that combined donepezil and CST can treat AD better than cholinesterase inhibitors alone. However, the level of improvement and the efficacy of CST are still somewhat debated. Future research is required to understand detailed interactions such as the effects of donepezil dosage (e.g., 5 mg, 10 mg, 23 mg) on combined therapies. It would also be beneficial to explore how different combinations of mixed interventions compare when treating AD (e.g., mixed donepezil, CST, and memantine treatment). Nonetheless, evidence suggests that CST is a decent treatment for AD, especially when combined with donepezil.

Combined donepezil and memantine treatment

When treating AD, it is also common for cholinesterase inhibitors to be combined with other drugs such as memantine. Memantine is known as an *N*-Methyl-D-aspartate receptor antagonists and works by regulating the neurotransmitter glutamate (Kishi et al., 2017). Glutamate regulation is important as the concentration of the neurotransmitter increases in AD, which can result in significant neuronal damage. In the literature, the combination of memantine and donepezil is the most common form of mixed AD treatment (Zemek et al., 2014). The combination of memantine and donepezil is also common in clinical practice (Alzheimer's Association, 2020). Accordingly, in this section, we discuss the efficacy and ability of combined donepezil and memantine drug treatments.

There is an evidence to suggest that combined memantine and donepezil treatment is better than individual drug treatments (e.g., donepezil alone). For example, a study by Cao et al. (2020) compared the therapeutic effects of donepezil when administered alone and when combined with memantine. The study specifically focused on investigating the efficacy and safety of each treatment for AD and chronic obstructive pulmonary disease. The data from approximately 310 participants were obtained from Hebei Medical University, where participants had received a baseline 10-mg dosage of donepezil. Cognitive and psychological measures were recorded using the MMSE, Dementia QOL score, Bristol Activities of Daily Living Scale, NPI, and the General Health Questionnaire. Cao et al. (2020) found that combined donepezil and memantine treatment was significantly better than just donepezil. Moreover, they found that there was no significant difference between groups when observing adverse effects.

In a meta-analysis on the topic, Chen et al. (2017) analyzed the findings of combined donepezil and memantine studies in the literature. Specifically, they sought to investigate the quality and findings of the combined therapy research in the literature. The articles for the meta-analysis were sourced from the Cochrane, PsycINFO, PubMed, Ovid

Medline, and Embase databases. From the 11 articles selected, participants cognition, psychology (e.g., confusion, emotions), behavior, global function, and adverse events were analyzed. Chen et al. (2017) found that a combined treatment was more effective than the administration of donepezil alone. Similarly, Cao et al. (2020) found that a combined memantine and donepezil treatment improved the cognitive, behavioral, psychological, and functional skills of an individual with AD. These effects also continued to improve the cognitive symptoms of AD during the later stages of the disease. Contrastingly, single cholinesterase inhibitor treatments often worsen at the later stages of AD progression. Chen et al. (2017) also found that there was no increased risk of adverse events when taking the combined treatment in comparison to donepezil alone.

In conclusion, evidence suggests that a mixed donepezil and memantine treatment can improve the neuropsychological symptoms of AD more so than individual treatments. Moreover, it has also been observed that a combined drug treatment is just as safe as donepezil when treating AD. Accordingly, researchers and clinicians should consider a combined donepezil and memantine treatment when possible. However, it is important to note that all complications (adverse effects) by consuming cholinesterase inhibitors still occur during mixed treatment. Subsequently, researchers and clinicians should observe individuals undergoing a combined drug therapy and adjust the treatment accordingly.

Conclusion

In this chapter, we assessed the use of donepezil for the treatment of AD. We specifically assessed donepezil due to its history and popularity as a cholinesterase inhibitor. Throughout this chapter, we explored the effects of dosage and the influence of the adverse effects of donepezil on its plausibility of use. We also sought to comment on combined treatments and directions for future research. We found that donepezil presents a mild to moderate improvement in the cognitive and functional symptoms of AD. This is supported by a sizable amount of the literature with little contention. However, the auxiliary effects of donepezil in AD (e.g., neurodegenerative, metabolic, and neuropsychiatric treatment) are more heavily debated.

In this chapter, we found that donepezil was observed to have a positive effect on the cognitive, functional, neurodegenerative (brain atrophy and volume), neuropsychiatric, and metabolic symptoms of AD.

However, the relationship between donepezil, atrophy and the neuropsychiatric symptoms of AD is rather preliminary and needs further research (due to conflicting results amongst studies). We also consistently found that the neuropsychological benefits of donepezil treatment increase with donepezil dosage (from 5 to 10 mg or 23 mg). However, dosage increase is also related to the occurrence and severity of adverse events such as diarrhea and nausea. To counteract the adverse effects of donepezil, most studies will either gradually transition individuals from a low to moderate dosage (e.g., titration from 5 to 10 mg) or use combined therapies (e.g., donepezil and memantine or CST) to supplement donepezil. Accordingly, the adverse effects of donepezil do not seem to outweigh the benefits of the drug when moderating dosage and using titration or combined treatment strategies. Nonetheless, there is still a need to further improve the adverse effects of donepezil due to its correlated effects with participants' QOL (Rogers et al., 1998).

One of the biggest problems with donepezil research is its handling of mortality rates. Many studies, especially those with higher doses, involve a small to moderate dropout of individuals due to their inability to deal with the adverse effects of the drug. Such participant mortality is problematic as the remaining participants in research are only those who respond well to the drug. Accordingly, an accurate representation and objective understanding of donepezil may be confounded by participant mortality. Future studies should identify those most susceptible to the adverse effects of donepezil and seek to moderate these risk factors. Researchers should also explore methods that decrease adverse events during donepezil treatment. The occurrence of participant mortality is closely related to the underlying problem of individual factors in donepezil research. Specifically, there are participant traits and environmental variables that affect an individual's treatment and drug acceptance. There has been some research that has shown the importance of genetic precedence and demographic traits in donepezil acceptance; however, these studies are new and lack deep research. Accordingly, researchers should seek to investigate these participant factors when conducting donepezil studies. It is also important that researchers closely watch and report statistical measures of validity when participant mortality occurs for the sake of transparency.

In conclusion, this chapter explores the efficacy of donepezil in the treatment of AD. We found that donepezil significantly improves factors of cognitive decline as well as showing evidence for affecting other factors of AD (e.g., neurodegeneration). This chapter was written to summarize the current literature and to direct and inform future research so that AD can be better treated and understood. In turn, we suggested that multiple areas require extensive research and replication. First, future research should explore the efficacy of combined treatments compared to high donepezil dosage treatments. Second, future research should investigate ways to control and decrease the adverse effects of donepezil treatments, especially in high dosage cohorts. Researchers should not be deterred from investigating high dosages of donepezil; however, they must seek to compensate for the increase in adverse effects when doing so. Third, future research should seek to further investigate the effects of individual variables on donepezil treatment quality and acceptance. Namely, researchers should investigate the effects of genetics, age, and nondemented illnesses such as diabetes on the quality and acceptance of donepezil treatment. These variables have been highlighted as crucial to treatment success yet lack research. Finally, research should study how the QOL of individuals with AD can be improved during donepezil treatment. We believe that individuals with AD should have the best treatment and QOL possible. As such, all these research areas align with this chapter's goal to better the lives of individuals with AD and increase the efficacy of donepezil.

References

Adlimoghaddam, A., Neuendorff, M., Roy, B., & Albensi, B. C. (2018). A review of clinical treatment considerations of donepezil in severe Alzheimer's disease. *CNS Neuroscience & Therapeutics*, *24*(10), 876–888. Available from https://doi.org/10.1111/cns.13035.

Alzheimer's Association. (2018). 2018 Alzheimer's disease facts and figures. *Alzheimer's & Dementia*, *14*(3), 367–429.

Alzheimer's Association. (2020). Medications for memory. *Alzheimer's Disease & Dementia. Retrieved from*. Available from https://alz.org/alzheimers-dementia/treatments/medications-for-memory.

Andersen, F., Viitanen, M., Halvorsen, D. S., Straume, B., Wilsgaard, T., & Engstad, T. A. (2012). The effect of stimulation therapy and donepezil on cognitive function in Alzheimer's disease. A community based RCT with a two-by-two factorial design. *BMC Neurology*, *12*, 59. Available from https://doi.org/10.1186/1471-2377-12-59.

Armstrong, R. A. (2013). What causes Alzheimer's disease? *Folia Neuropathologica*, *51*(3), 169–188. Available from https://doi.org/10.5114/fn.2013.37702.

Bartus, R. T., Dean, R. L., Beer, B., & Lippa, A. S. (1982). The cholinergic hypothesis of geriatric memory dysfunction. *Science (New York, N.Y.)*, *217*(4558), 408–414. Available from https://doi.org/10.1126/science.7046051.

Benjamin, B., & Burns, A. (2007). Donepezil for Alzheimer's disease. *Expert Review of Neurotherapeutics*, 7(10), 1243–1249. Available from https://doi.org/10.1586/14737175.7.10.1243.

Birks, J., & Harvey, R. J. (2006). Donepezil for dementia due to Alzheimer's disease. *The Cochrane Database of Systematic Reviews* (1), CD001190. Available from https://doi.org/10.1002/14651858.CD001190.pub2.

Blanco-Silvente, L., Castells, X., Saez, M., Barceló, M. A., Garre-Olmo, J., Vilalta-Franch, J., & Capellà, D. (2017). Discontinuation, efficacy, and safety of cholinesterase inhibitors for Alzheimer's disease: A meta-analysis and meta-regression of 43 randomized clinical trials enrolling 16 106 patients. *The International Journal of Neuropsychopharmacology*, *20*(7), 519–528. Available from https://doi.org/10.1093/ijnp/pyx012.

Cacabelos, R. (2007). Donepezil in Alzheimer's disease: From conventional trials to pharmacogenetics. *Neuropsychiatric Disease and Treatment*, *3*(3), 303–333.

Cammisuli, D. M., Danti, S., Bosinelli, F., & Cipriani, G. (2016). Non-pharmacological interventions for people with Alzheimer's Disease: A critical review of the scientific literature from the last ten years. *European Geriatric Medicine*, 7(1), 57–64. Available from https://doi.org/10.1016/j.eurger.2016.01.002.

Cao, Y., Qian, L., Yu, W., Li, T., Mao, S., & Han, G. (2020). Donepezil plus memantine versus donepezil alone for treatment of concomitant Alzheimer's disease and chronic obstructive pulmonary disease: A retrospective observational study. *Journal of International Medical Research*, *48*(2). Available from https://doi.org/10.1177/0300060520902895.

Chang, Y.-P., Yang, C.-H., Chou, M.-C., Chen, C.-H., & Yang, Y.-H. (2015). Clinical compliance of donepezil in treating Alzheimer's disease in Taiwan. *American Journal of Alzheimer's Disease & Other Dementias*, *30*(4), 346–351. Available from https://doi.org/10.1177/1533317514556875.

Chapman, S. B., Weiner, M. F., Rackley, A., Hynan, L. S., & Zientz, J. (2004). Effects of cognitive-communication stimulation for Alzheimer's disease patients treated with donepezil. *Journal of Speech, Language, and Hearing Research*, 47(5), 1149–1163.

Chen, J., Duan, Y., Li, H., Lu, L., Liu, J., & Tang, C. (2019). Different durations of cognitive stimulation therapy for Alzheimer's disease: A systematic review and meta-analysis. *Clinical Interventions in Aging*, *14*, 1243–1254. Available from https://doi.org/10.2147/CIA.S210062.

Chen, R., Chan, P.-T., Chu, H., Lin, Y.-C., Chang, P.-C., Chen, C.-Y., & Chou, K.-R. (2017). Treatment effects between monotherapy of donepezil versus combination with memantine for Alzheimer disease: A meta-analysis. *PLoS One*, *12*(8), e0183586. Available from https://doi.org/10.1371/journal.pone.0183586.

Craig, L. A., Hong, N. S., & McDonald, R. J. (2011). Revisiting the cholinergic hypothesis in the development of Alzheimer's disease. *Neuroscience & Biobehavioral Reviews*, *35*(6), 1397–1409. Available from https://doi.org/10.1016/j.neubiorev.2011.03.001.

Cummings, J., Lai, T. -J., Hemrungrojn, S., Mohandas, E., Kim, S. Y., Nair, G., & Dash, A. (2016). Role of donepezil in the management of neuropsychiatric symptoms in Alzheimer's disease and dementia with Lewy bodies. *CNS Neuroscience & Therapeutics*, *22*(3), 159–166. Available from https://doi.org/10.1111/cns.12484.

Davidson, P. S. R., & Winocur, G. (2010). Aging and cognition. In G. F. Koob, M. L. Moal, & R. F. Thompson (Eds.), *Encyclopedia of behavioral neuroscience* (pp. 20–26). London: Academic Press. Available from https://doi.org/10.1016/B978-0-08-045396-5.00031-2.

Doody, R. S., Geldmacher, D. S., Farlow, M. R., Sun, Y., Moline, M., & Mackell, J. (2012). Efficacy and safety of donepezil 23 mg versus donepezil 10 mg for moderate-to-severe Alzheimer's disease: A subgroup analysis in patients already taking or not taking concomitant memantine. *Dementia and Geriatric Cognitive Disorders*, *33*(2–3), 164–173. Available from https://doi.org/10.1159/000338236.

Dubois, B., Chupin, M., Hampel, H., Lista, S., Cavedo, E., Croisile, B., … Dormont, D. (2015). Donepezil decreases annual rate of hippocampal atrophy in suspected prodromal Alzheimer's disease. *Alzheimer's & Dementia*, *11*(9), 1041–1049. Available from https://doi.org/10.1016/j.jalz.2014.10.003.

Espiritu, A. I., & Cenina, A. R. F. (2020). The effectiveness and tolerability of the high dose donepezil at 23 mg tablet per day for Alzheimer's disease: A meta-analysis of randomized controlled trials. *Acta Medica Philippina, 54*(3), Article 3. Retrieved from. Available from https://actamedicaphilippina.upm.edu.ph/index.php/acta/article/view/1669.

Farlow, M. R., Salloway, S., Tariot, P. N., Yardley, J., Moline, M. L., Wang, Q., ... Satlin, A. (2010). Effectiveness and tolerability of high-dose (23 mg/d) versus standard-dose (10 mg/d) donepezil in moderate to severe Alzheimer's disease: A 24-week, randomized, double-blind study. *Clinical Therapeutics, 32*(7), 1234–1251. Available from https://doi.org/10.1016/j.clinthera.2010.06.019.

Ferreira-Vieira, T. H., Guimaraes, I. M., Silva, F. R., & Ribeiro, F. M. (2016). Alzheimer's disease: Targeting the cholinergic system. *Current Neuropharmacology, 14*(1), 101–115. Available from https://doi.org/10.2174/1570159X13666150716165726.

Francis, P. T., Palmer, A. M., Snape, M., & Wilcock, G. K. (1999). The cholinergic hypothesis of Alzheimer's disease: A review of progress. *Journal of Neurology, Neurosurgery & Psychiatry, 66*(2), 137–147. Available from https://doi.org/10.1136/jnnp.66.2.137.

Griffith, P., Lichtenberg, P., Goldman, R., & Payne-Parrish, J. (2006). Safety and efficacy of donepezil in African Americans with mild-to-moderate Alzheimer's disease. *Journal of the National Medical Association, 98*(10), 1590–1597.

Haake, A., Nguyen, K., Friedman, L., Chakkamparambil, B., & Grossberg, G. T. (2020). An update on the utility and safety of cholinesterase inhibitors for the treatment of Alzheimer's disease. *Expert Opinion on Drug Safety, 19*(2), 147–157. Available from https://doi.org/10.1080/14740338.2020.1721456.

Ishiwata, A., Mizumura, S., Mishina, M., Yamazaki, M., & Katayama, Y. (2014). The potentially protective effect of donepezil in Alzheimer's disease. *Dementia and Geriatric Cognitive Disorders, 38*(3–4), 170–177. Available from https://doi.org/10.1159/000358510.

Jackson, S., Ham, R. J., & Wilkinson, D. (2004). The safety and tolerability of donepezil in patients with Alzheimer's disease. *British Journal of Clinical Pharmacology, 58*(Suppl. 1), 1–8. Available from https://doi.org/10.1111/j.1365-2125.2004.01848.x.

Jia, J., Wei, C., Jia, L., Tang, Y., Liang, J., Zhou, A., ... Doody, R. S. (2017). Efficacy and safety of donepezil in Chinese patients with severe Alzheimer's disease: A randomized controlled trial. *Journal of Alzheimer's Disease, 56*(4), 1495–1504. Available from https://doi.org/10.3233/JAD-161117.

Kim, K., Han, J. W., So, Y., Seo, J., Kim, Y. J., Park, J. H., ... Kim, K. W. (2017). Cognitive stimulation as a therapeutic modality for dementia: A meta-analysis. *Psychiatry Investigation, 14*(5), 626–639. Available from https://doi.org/10.4306/pi.2017.14.5.626.

Kishi, T., Matsunaga, S., Oya, K., Nomura, I., Ikuta, T., & Iwata, N. (2017). Memantine for Alzheimer's disease: An updated systematic review and meta-analysis. *Journal of Alzheimer's Disease, 60*(2), 401–425. Available from https://doi.org/10.3233/JAD-170424.

Kobayashi, H., Ohnishi, T., Nakagawa, R., & Yoshizawa, K. (2016). The comparative efficacy and safety of cholinesterase inhibitors in patients with mild-to-moderate Alzheimer's disease: A Bayesian network meta-analysis. *International Journal of Geriatric Psychiatry, 31*(8), 892–904. Available from https://doi.org/10.1002/gps.4405.

Lu, J., Wan, L., Zhong, Y., Yu, Q., Han, Y., Chen, P., ... Guo, C. (2015). Stereoselective metabolism of donepezil and steady-state plasma concentrations of S-donepezil based on CYP2D6 polymorphisms in the therapeutic responses of Han Chinese patients with Alzheimer's disease. *Journal of Pharmacological Sciences, 129*(3), 188–195. Available from https://doi.org/10.1016/j.jphs.2015.10.010.

Ma, Y., Ji, J., Li, G., Yang, S., & Pan, S. (2018). Effects of donepezil on cognitive functions and the expression level of β-amyloid in peripheral blood of patients with Alzheimer's disease. *Experimental and Therapeutic Medicine*, *15*(2), 1875–1878.

Matsuda, O. (2007). Cognitive stimulation therapy for Alzheimer's disease: The effect of cognitive stimulation therapy on the progression of mild Alzheimer's disease in patients treated with donepezil. *International Psychogeriatrics*, *19*(2), 241–252.

Moon, C.-M., Kim, B.-C., & Jeong, G.-W. (2016). Effects of donepezil on brain morphometric and metabolic changes in patients with Alzheimer's disease: A DARTEL-based VBM and 1H-MRS. *Magnetic Resonance Imaging*, *34*(7), 1008–1016. Available from https://doi.org/10.1016/j.mri.2016.04.025.

Moss, D. E. (2020). Improving anti-neurodegenerative benefits of acetylcholinesterase inhibitors in Alzheimer's disease: Are irreversible inhibitors the future? *International Journal of Molecular Sciences*, *21*(10), 3438. Available from https://doi.org/10.3390/ijms21103438.

Pink, J., O'Brien, J., Robinson, L., & Longson, D. (2018). Dementia: Assessment, management and support: Summary of updated NICE guidance. *BMJ (Clinical Research Edition)*, *361*, *k2438*. Available from https://doi.org/10.1136/bmj.k2438.

Rogers, S. L., Farlow, M. R., Doody, R. S., Mohs, R., & Friedhoff, L. T. (1998). A 24-week, double-blind, placebo-controlled trial of donepezil in patients with Alzheimer's disease. Donepezil Study Group. *Neurology*, *50*(1), 136–145. Available from https://doi.org/10.1212/wnl.50.1.136.

Sharma, K. (2019). Cholinesterase inhibitors as Alzheimer's therapeutics (Review). *Molecular Medicine Reports*, *20*(2), 1479–1487. Available from https://doi.org/10.3892/mmr.2019.10374.

Singh, R., & Sadiq, N. M. (2020). *Cholinesterase inhibitors*. StatPearls Publishing. Available from http://www.ncbi.nlm.nih.gov/books/NBK544336/.

Teipel, S. J., Grinberg, L. T., Hampel, H., & Heinsen, H. (2009). Cholinergic system imaging in the healthy aging process and Alzheimer disease. In L. R. Squire (Ed.), *Encyclopedia of neuroscience* (pp. 857–868). London: Academic Press. Available from https://doi.org/10.1016/B978-008045046-9.02041-6.

Wang, L., Harms, M. P., Staggs, J. M., Xiong, C., Morris, J. C., Csernansky, J. G., & Galvin, J. E. (2010). Donepezil treatment and changes in hippocampal structure in very mild Alzheimer's disease. *Archives of Neurology*, *67*(1), 99–106. Available from https://doi.org/10.1001/archneurol.2009.292.

Yatabe, Y., Hashimoto, M., Kaneda, K., Honda, K., Ogawa, Y., Yuuki, S., & Ikeda, M. (2013). Efficacy of increasing donepezil in mild to moderate Alzheimer's disease patients who show a diminished response to 5 mg donepezil: A preliminary study. *Psychogeriatrics: The Official Journal of the Japanese Psychogeriatric Society*, *13*(2), 88–93. Available from https://doi.org/10.1111/psyg.12004.

Zemek, F., Drtinova, L., Nepovimova, E., Sepsova, V., Korabecny, J., Klimes, J., & Kuca, K. (2014). Outcomes of Alzheimer's disease therapy with acetylcholinesterase inhibitors and memantine. *Expert Opinion on Drug Safety*, *13*(6), 759–774. Available from https://doi.org/10.1517/14740338.2014.914168.

Zhang, X., Yu, R., Wang, H., & Zheng, R. (2020). Effects of rivastigmine hydrogen tartrate and donepezil hydrochloride on the cognitive function and mental behavior of patients with Alzheimer's disease. *Experimental and Therapeutic Medicine*, *20*(2), 1789–1795. Available from https://doi.org/10.3892/etm.2020.8872.

Index

Note: Page numbers followed by "*f*" and "*t*" refer to figures and tables, respectively.

FSC
www.fsc.org
MIX
Papier | Fördert
gute Waldnutzung
FSC® C083411

Kolibri 360 GmbH
Ferdinand-Jühlke-Straße 7
99095 Erfurt, Deutschland
produktsicherheit@kolibri360.de